Fighting Global Neo-Extractivism

I0131713

Fighting Global Neo-Extractivism: Fossil-Free Social Movements in South Africa analyses social struggles over damaging new fossil-fuel projects in the Global South with a focus on South Africa, Africa's biggest fossil fuel emitter.

Fossil-fuel extraction in South Africa has reached a new accelerated phase in which the fossil-fuel frontier is moving beyond historical 'sacrifice zones' into non-traditional spaces, such as conservation parks and middle-class neighbourhoods, and provoking fervent opposition from grassroots activists. This book examines campaigns such as Frack Free South Africa and Save our iMfolozi Wilderness, viewing them as struggles against neo-extractivism driven by the state and industry. Through a series of detailed case studies, it highlights the shaping of mobilisation patterns by prior land use practices and the capacity to mobilise different social groups across race and class. Developing the notion of the fossil-fuel frontier as the material and political boundary that activists in South Africa and elsewhere in the world render visible, this volume provides a theoretical framework to understanding global mobilisation patterns.

This timely and impassioned book will appeal to students and researchers interested in a range of subjects, including environmentalism, social movements, political ecology, and development studies.

Jasper Finkeldey is Postdoctoral Researcher and Lecturer in Political Science in the Faculty of Social Sciences and Cultural Studies at the Martin Luther University of Halle-Wittenberg, Germany. His interdisciplinary research focuses on environmental politics, social movements, political economy, and conflicts over resources.

The *Mobilization* Series on Social Movements, Protest, and Culture

Series editor: Professor Hank Johnston
San Diego State University, USA

Published in conjunction with *Mobilization: An International Quarterly*, the premier research journal in the field, this series publishes a broad range of research in social movements, protest and contentious politics. This is a growing field of social science research that spans sociology and political science as well as anthropology, geography, communications and social psychology. Enjoying a broad remit, the series welcome works on the following topics: social movement networks; social movements in the global South; social movements, protest, and culture; personalist politics, such as living environmentalism, guerrilla gardens, anticonsumerist communities, anarchist-punk collectives; and emergent repertoires of contention.

Exiled Activism
Political Mobilization in Egypt and England
David McKeever

Resisting the Backlash
Street Protest in Italy
Donatella della Porta, Niccolò Bertuzzi, Daniela Chironi, Chiara Milan, Martín Portos and Lorenzo Zamponi

Fukushima and Civil Society
The Japanese Anti-Nuclear Movement from a Socio-Political Perspective
Beata Bochorodycz

Fighting Global Neo-Extractivism
Fossil-Free Social Movements in South Africa
Jasper Finkeldey

For more information about this series, please visit: www.routledge.com/The-Mobilization-Series-on-Social-Movements-Protest-and-Culture/book-series/ASHSER1345

Fighting Global Neo-Extractivism

Fossil-Free Social Movements in South Africa

Jasper Finkeldey

Routledge
Taylor & Francis Group
LONDON AND NEW YORK

First published 2023
by Routledge
4 Park Square, Milton Park, Abingdon, Oxon OX14 4RN

and by Routledge
605 Third Avenue, New York, NY 10158

Routledge is an imprint of the Taylor & Francis Group, an Informa business

British Library Cataloguing-in-Publication Data
A catalogue record for this book is available from the British Library

Library of Congress Cataloging-in-Publication Data
A catalog record for this book has been requested

ISBN: 978-0-367-62012-7 (hbk)
ISBN: 978-0-367-62796-6 (pbk)
ISBN: 978-1-003-11083-5 (ebk)

DOI: 10.4324/9781003110835

Typeset in Times New Roman
by Apex CoVantage, LLC

To my family and friends

Contents

PART II
Social Movements Fighting Fossil Fuels   53

Figures

Tables

Abbreviations

ACDP	African Christian Democratic Party
ANC	African National Congress
CBO	Community-based organisation
CCS	Centre for Civil Society
CDU	Christian Democratic Party of Germany
CER	Centre for Environmental Rights
COSATU	Congress of South African Trade Unions
DA	Democratic Alliance
DMR	Department of Mineral Resources
DoE	Department of Energy
EIUG	Energy Intensive User Group
EJ	Environmental Justice
EU	European Union
FFF	Fossil Fuel Foundation of Africa
FFSA	Frack Free South Africa
GDR	German Democratic Republic
GET	Global Environmental Trust
GND	Green New Deal
I&AP	Interested and affected parties
IEC	Electoral Commission of South Africa
ITB	Ingonyama Trust Board
KZN	KwaZulu-Natal
MEC	Minerals-Energy Complex
NGO	Non-governmental organisation
RBCT	Richards Bay coal terminal
SABC	South African Broadcasting Corporation
SACP	South African Communist Party
SAVE	Save Our iMfolozi Wilderness Campaign
SMO	Social movement organisation
SPD	Social Democratic Party of Germany
TKAG	Treasure the Karoo Action Group
WSF	World Social Forum
WTO	World Trade Organization
ZAC	Zululand Anthracite Colliery

Acknowledgements

This research truly feels like a journey.

This research is on South Africa, where I met a number of very inspiring people. I thank Gcina Phakamile Makoba for her assistance and critical eye.

I thank Adrian Nel, Patrick Bond, Shauna Mottiar, and Sarah Bracking for welcoming me at the University of KwaZulu-Natal and providing me with the most stimulating academic environment. Your help went far beyond expectations.

I also thank the tireless activists and my comrades Sheila Berry, Rob Symons, and Billy. I am indebted to you for everything you taught me about activism. Bobby Peek, Desmond D'Sa, and Sifiso Dladla also greatly enhanced my understanding of what it takes to be a full-time activist. Thanks a lot to Jenny Longmore at Ezemvelo Wildlife for her generous help to carry out this research.

Thanks to Dumisani Nkabinde for your endless patience and friendship. Without Dumisani, I would only have learned half of what I did about South Africa. I feel lucky to call you my friend.

Thanks to Leo for our eye-opening encounters and your generosity. Thanks to Xola for the lovely time we passed philosophising on life.

I thank the Bux and Watson families. This is where it all started.

I thank my friend Mdu for his light-heartedness and uncompromising political analysis.

I thank my hosts at Glenmore Pastoral Centre.

I also thank my friend Sihlangu, who taught me a lot about Newcastle by day and night.

I thank Bernadette, who was the most wonderful and resourceful host in Durban.

Thanks to my supervisors Jane Hindley and Steffen Böhm, who were always there when I needed them. I count myself lucky to have worked with you. Thanks to the support of Diane Holt, who made an unexpected return to South Africa possible. Thanks to Melissa Tyler, Stevphen Shukaitis, and Casper Hoedemaekers, who kept their keen eyes on my work as chairs. Thanks also to Essex Business School for providing the space and resources. Thanks to my friends at EBS who have lit up my stay at Essex, and my Exeter friends who made me feel most welcome when I came to visit. And finally, thanks to the Economic and Social Research Council for providing generous funding for this research. This journey would not have been possible without this material base.

Thanks to my parents who never stop believing in me. Thanks to my brother Tim for his loyalty and support. Thanks to Susanne who is a curious reader of my work and encouraged me throughout the process. Thanks to my former students at Freie Universität Berlin who made my academic work feel meaningful and exciting. Thanks to Jakob, who paid a memorable visit to South Africa and remains a loving and supportive friend. Thanks to Torben Fischer, who has been a true friend and an amazing colleague. Thanks to Hendrik Theine, who is my oldest friend and partner in PhD trials and tribulations. Thanks to Magherita and to Clara who have been integral to this journey even though our paths split. Thanks to the Centre for Management Research at the University Paris-Dauphine where I was honoured to take up my first academic position. Thanks to Petra Dobner at the Martin Luther University of Halle-Wittenberg for her support and trust and the team at the Chair of System Analysis and Comparative Politics. Thanks to Katharina for your help in the final stages. Finally to Maja who has just seen the light and brings so much joy.

This research is written in memory of late Ulrike von Dellingshausen. Your attitude towards life was an example to those who had the pleasure of meeting you. We miss you a lot.

Halle (Saale), May 2022

1 Introduction

For communities living in the vicinity of fossil-fuel infrastructure, toxic leaks, displacement, and explosions have been a reality from the inception of the fossil-fuel era. 'Sacrifice zones' have since sprung up in the parts of the world where fossil deposits are opened for exploitation (Klein 2015). Even though the fossil-fuel industry has tried to keep it secret for a long time, the burning of fossil fuels is a major driver of climate change (Oreskes and Conway 2012). The Intergovernmental Panel on Climate Change (IPCC) estimates that between 1970 and 2010, 78% of CO_2 emissions came from fossil fuels and industry (Blanco *et al*. 2014, p. 354). However, even after the Paris Climate Agreement in late 2015, the 20 largest economies in the world (G20) provided three times more funds to fossil fuels than to renewable energy (Oil Change International and Friends of the Earth 2020). As it stands, the post-Covid-19 economic recovery will largely be fossil-fuelled in the foreseeable future. But this book is less about predictions and more about contemporary questions of how South African and global fossil-free social movements challenge and change political structures on a warming planet. Therefore we need a thorough understanding of the political power structures enabling fossil fuels.

The fossil-fuel industry remains powerful in financial strength and political connection (Goodman *et al*. 2020). Fossil-fuel interests are deeply entrenched in the fossil-fuel-producing states and the world economy (Di Muzio 2015; Mitchell 2013). This is most obvious for state-owned enterprises, such as Petrobras in Brazil, Statoil in Norway, or the Saudi Arabian Oil Company: in these countries especially fossil-fuel interests are *raison d'État*. In emerging economies like Brazil, Indonesia and South Africa, new fossil-fuel infrastructure is under construction and presented as an integral part of the development model. The global supply chains are dripping with fossil fuels. These factors result in the runaway extraction and consumption of fossil fuels globally (see Figure 1.1).

However, today's political juncture leaves the 'fossil-fuel bloc' increasingly isolated, and 'substantial possibilities for change have begun to emerge' (Goodman *et al*. 2020, p. 41). Oil wells, gas pipelines, and coal mines increasingly symbolise an uneasiness with the destruction of the planet's climate and species extinction. Demonstrations in major fossil-fuel developing countries, such as

DOI: 10.4324/9781003110835-1

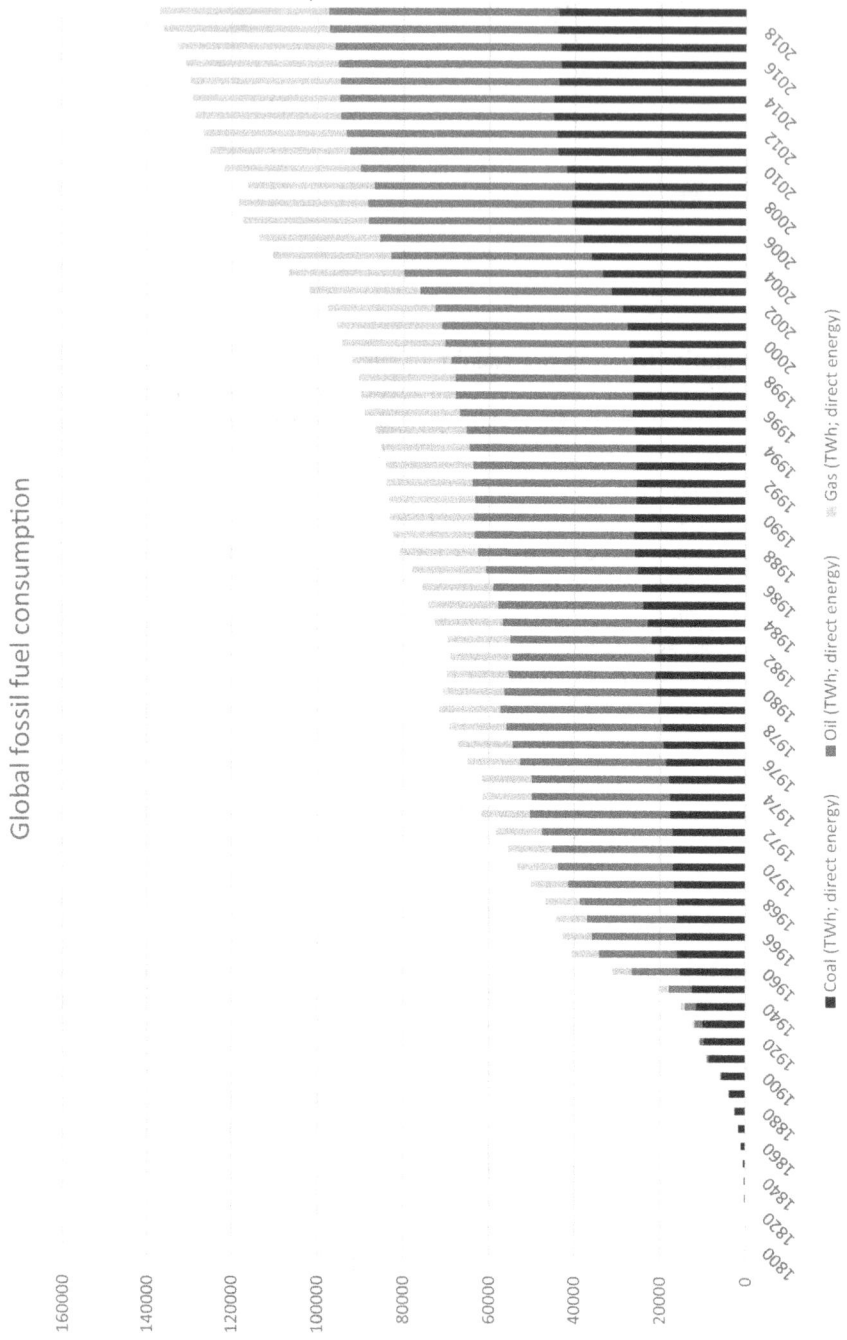

Figure 1.1 Global fossil fuel consumption.

South Africa, the USA, Canada, Germany, and Australia, are making headlines. Banners flying at demonstrations to 'end dirty coal' or to 'frack off' and blocking coal mines and pipelines pressure governments to reconsider whether they should allow the fossil deposits buried underground to be exploited. While the fossil-fuel era is everything but history, new social movements emerge, as fossil-fuel interests are still promoted and sustained (de Graaf and Sovacool 2020, pp. 133–139). Activists draw lines between their livelihoods and fossil-fuel projects. Anti-fossil-fuel movements challenge fossil-fuel exploitation and investment by state and private actors. They do so by raising awareness for the livelihoods that are destroyed locally and pointing out how the burning of fossil fuels accelerates climate change globally.

Social movements are more and more recognised for actively shaping energy policy (Temper *et al.* 2020; Piggot 2018). I define social movements against fossil fuels or fossil-free movements (which I will use interchangeably) as organised and sustained groups using different tactics to stop fossil-fuel projects at the point of extraction or any other point of the production cycle. Fossil-free activists are at particular risk of violence, as they threaten the profits of a billion-dollar industry that is often backed by the state and therefore heavily policed. The challenge to stop fossil fuels might be coupled with a range of other demands and frames around conservation, climate change, or anti-racism.

How struggles against fossil fuels play out is context-specific and may involve diverse people and organisations, mainly in fossil-fuel-producing states. Climate activists put increasing pressure on governments and the fossil-fuel industry to bring the fossil-fuel era to an end. The Fridays for Future movement in Europe, Frack Free groups around the world, the Sunrise Movement in the USA, Extinction Rebellion in Western Europe, and the Standing Rock mobilisation against the Keystone XL pipeline from Canada to the USA made claims against the fossil-fuel industry and mobilised impressive numbers of activists over time. Social movement mobilisations against fossil fuels shift the climate change debate and influence fossil fuel use (Temper *et al.* 2020).

Mobilisations against fossil-fuel projects are often happening in countries with heavy carbon footprints outside the Western hemisphere. The fight of the Ogoni people in the Niger Delta or the Yasunì-ITT initiative in Ecuador against oil is among the better-known examples. But most of the global struggles against fossil fuels are untold. The names of activists involved and grievances are never voiced. These 'grassroots environmentalists' (Staggenborg 2020) are an integral part of the global movement changing energy politics from the bottom up. Often, they form alliances with bigger movements against fossil fuels.

This book explores struggles against fossil fuels in South Africa, Africa's biggest coal producer and CO_2 emitter from fossil fuels. As I will show in this book, the minerals-energy complex (MEC) – understood as the tight interlinkages among South Africa's mining and energy sectors, finance, and the state (Fine 2008; Fine and Rustomjee 1996) – is alive and kicking. However, the South African fossil-fuel bloc is challenged by social movements from within the country and faces international pressure for de-carbonisation, emanating from international climate treaties.

In 2017, 90% of the coal consumed in Africa was produced in South Africa (Mining Africa 2017). As part of the government's growth strategy to expand onshore and offshore fossil-fuel production, multiple sites are currently under application, including lands bordering residential, farm, and conservation areas (Petroleum Agency South Africa n.d.). Existing historical studies provide important insights into understanding fossil-fuel expansionism in South Africa (Hallowes 2011; Lang 1995). However, contemporary fossil-fuel expansion and resistance sites have not yet been studied in any great detail in South Africa. While empirical studies on high-profile anti-fossil-fuel resistances emphasise dispossession, violence, criminalisation of activists, and resistance elsewhere (e.g. Sander 2017; Coryat and Picq 2016; Okonta and Oronto 2003), little has been written about contentious fossil-fuel frontiers in South Africa (useful exceptions include Bond 2018b; Leonard 2016, 2011).

This book offers both an analysis of fossil-fuel hegemony defined as political power entrenching fossil fuels and social movement claim-making against new fossil-fuel projects in South Africa. Fossil-fuel hegemony is place-based and takes different forms depending on previous land uses, forms of resistance, and claim-making of anti-fossil-fuel movements. I offer a qualitative analysis of two different social movement campaigns in the South African province of KwaZulu-Natal (KZN) informed by fieldwork undertaken on the ground. Formed in 2014, the Save our iMfolozi Wilderness campaign's (SAVE) goal is to stop coal mining expansion at the border of the oldest African nature reserve, Hluhluwe–iMfolozi Park. SAVE's campaign is led by two activists of the Global Environmental Trust (GET) who built and extended a social movement network and organise a social movement campaign. Under the SAVE umbrella, GET managed to bring together 11 social and environmental groups determined to stop coal-mining projects in the area. The other social movement campaign I research is Frack Free South Africa (FFSA). This alliance began in 2015 with core activists recruited from the KZN Midlands, which is an affluent part of the province. Many nature conservancies and local businesses in areas proposed for fracking support FFSA. Both SAVE and FFSA stage protests at public consultation meetings and come up with slogans to defend places they value. Their ongoing efforts to stop fossil-fuel expansion have also been covered in the press. None of the campaigns is officially supported by larger organisations such as unions or political parties. This lack of support makes it particularly challenging for FFSA and SAVE to access resources and sustain their campaigns. As I will show, even within KwaZulu-Natal province, the differences in actor configurations and prior land use shape the scope and intensity of anti-fossil-fuel mobilisations.

I will also show that the involvement of NGOs is a key in starting and sustaining anti-fossil-fuel mobilisation. NGOs are assuming a leadership role in anti-fossil-fuel mobilisation in South Africa. Today's resource conflict frontiers can no longer be described as conflicts between labour and capital. The number of actors involved in mining conflicts has increased to include 'new NGO alliances, activist scholars, overall more international political engagement', who have changed the face of anti-mining struggles (Bebbington *et al.* 2008, p. 902). Increasingly, fossil-fuel projects are resisted before the commencement of operations.

Neo-Extractivism, Fossil Fuels, and the Climate Crisis in South Africa

In African contexts and elsewhere, states have embarked on a 'full-fledged neo-liberalism' path from the 1990s onwards. Major changes in the macroeconomic environment and structural adjustment policies have precipitated changes in fiscal policy, which have also led to the decline of formal sector jobs in industries such as mining (Larmer and Laterza 2017, p. 703; Larmer 2010). The tendency to destroy the 'forces of production' by wrecking livelihoods and poor working conditions under neo-liberal capitalism simultaneously provoked powerful resistance movements in Africa and elsewhere (Petras and Veltmeyer 2014, p. 20).

In the Global South, the extension of extractivism in remote areas is disproportionately affecting indigenous peoples and small farmers (Petras and Veltmeyer 2014, p. 1). The neo-liberal extraction model in the Global South is marked by an ongoing 'growth and export-oriented development strategy' driven mainly by liberalisation of the resources sectors (Engels and Dietz 2017, p. 1). Critical discussions of the global neoliberal extraction model focus increasingly on questions around climate change and environmental degradation (Pirani 2018; Malm 2016; Di Muzio 2015).

South Africa's government faces contradictions between its promotion of fossil-fuel extractivism and the commitments made at the Paris Climate Summit. In the context of the Paris summit, the South African government has recognised that the burning of fossil fuels results in 'marked temperature increases, rainfall variation, and rising sea levels as well as an increased frequency of severe weather events' (UNFCCC 2016). However, the government's projected industrial policy conflicts with the pledge of reaching peak emission between 2025 and 2030 and reducing its emission footprint thereafter (Climate Action Tracker 2018).

Coal remains the primary source for energy generation in South Africa with an over 90% contribution to the grid, followed by nuclear, which adds just over 5% (World Bank 2018; Department of Energy 2016). As a signatory to the Paris Agreement, South Africa has recognised adverse impacts stemming from climate change. Reportedly 71% of greenhouse gas (GHG) emissions since 1988 originate from fossil-fuel-producing companies (Griffin 2017). South Africa already suffers from climate change-related droughts and other extreme weather events (Hallowes 2015). From the 1970s onwards, South Africa's state-owned utility Eskom has massively increased coal input, resulting in a rise in the power grid's contribution to climate change (Burton and Winkler 2014, p. 2). South Africa's carbon dioxide emissions from energy consumption are the 12th highest worldwide (The Guardian Datablog 2016).

According to Bond (2008, p. 1047), climate change activism in South Africa is vital, because 'CO_2 emissions, measured as a percentage of per capita GDP, are twenty times higher than even those of the United States', thanks to the economy's excessive reliance on coal. Activists are questioning the current neo-liberal model of extractivism that underlies the imminent fossil-fuel expansion. South Africa's high numbers of protests can also be attributed to the intensification of environmental crises and immanent land grabs (Runciman *et al.* 2016; Bond and Mottiar 2013). Tensions arising from these interlinking crises have caused several casualties in South Africa's recent history. While the Marikana mine workers' struggle was forcefully thwarted, leaving 34 workers dead on 16 August 2012 (Alexander *et al.* 2012), an anti-mining activist lost life to environmental activism in the Eastern Cape in March 2016. Activists in KZN and elsewhere are repeatedly abused as 'anti-national, anti-people, anti-development' (Yeld 2018).

Neo-liberal extractivism in South Africa leaves a huge environmental footprint in mine-hosting communities and often ignores its own social and environmental responsibility targets (Forslund and Etkind 2013). As Klein (2015) writes, mine-hosting communities become 'sacrifice zones' in the form of 'black lungs for the coal miners or the poisonous waterways surrounding the mines' (pp. 172–173). Growing alienation from the existing extraction model can be seen in the emergence of anti-extractive social movements.

The geographical focus of this study is KZN province. I use KZN located in the southeast of South Africa as a case study, because it exemplifies the new fossil-fuel expansionism in South Africa. Socio-environmental struggles against polluting industries, including petrochemical industries in Durban and coal mining in Somkhele, have previously shown that social movements are facing up to state-backed industrial projects (KaManzi and Bond 2015; Leonard and Pelling 2010). KZN currently has only four high-grade coal mines operating (Eberhard 2011, p. 3). While historically being a labour-providing province to the mining industry (van Wyk 2013), there are plans to open estimated fossil-fuel reserves in the subsurface of the province. Close to 40% of the province is earmarked for gas exploration (likely via 'hydraulic fracturing' or 'fracking') (Savides 2017). There are also applications for controversial high-grade coal projects in the proximity of Africa's oldest national reserve.

One of the largest export terminals in the world is situated in KZN. The Richards Bay coal terminal (RBCT) services clients in both the east (India and China were the biggest customers between 2010 and 2017) and the west (where Spain and Italy were the biggest customers in the same timeframe). RBCT ships out more than 98% of coal, with just over 1% leaving the country from Durban (South African Market Insights 2017). Therefore, coal exports are almost exclusively leaving the country from KZN. Proximity to the export infrastructure is a crucial asset for companies aiming to open new mines and export coal.

Current applications for fossil-fuel expansion can radically shift the social and economic fabric of KZN province. As I will show, coal mining was introduced in the 1880s as a symbol of progress, helping to cheaply electrify households (Davenport 2013; Singer 2011). Pollution from coal was portrayed as a necessary by-product. With the latest introduction of new cleaner coal-fired power stations in South Africa, there is a newfound belief that fossil fuels are compatible with human and environmental health. Another argument in favour of coal is its abundance and low production costs.

This book looks closely at how new fossil-fuel projects are part of a broader hegemonic accumulation regime and discourse to expand the fossil-fuel frontier.

Powerful corporate actors entrench political consensus that social movement campaigns challenge in places they value. In places with historical geography that privileged conservation or in white neighbourhoods, anti-fossil-fuel project mobilisation is particularly persistent. In the absence of political parties supporting large-scale renewable transitions, social movement campaigns set the limits of fossil-fuel extraction in South Africa.

My case study consists of two social movement campaigns using 'thick' descriptions to make sense of the 'bounded reality' (Snow and Trom 2002) of proposed fossil-fuel projects in KZN. Using a theoretical framework building on the Italian intellectual Antonio Gramsci, I explain that resistance relies on 'organic intellectuals' weaving together disparate grievances against fossil fuels. Unless opposition to fossil fuels can unite, no political frontier against fossil mining can be established. Conservationists, local business people, and residents need to join forces to voice their grievances against fossil-fuel mining. My research shows that NGOs play an important role in articulating and coordinating anti-extractivist struggles. Moreover, NGOs create opportunities to increase and maintain the resistance against fossil-fuel projects in KZN.

Post-Apartheid State and Social Movements

Movements and their intellectual leaders need to be studied by considering the 'general complex of social relations' they are embedded in (Gramsci 1991, p. 8). Democratic civil society organisers under apartheid were united in their opposition to the racist apartheid government of the National Party (NP). The United Democratic Front (UDF) that gained momentum in the 1980s was formed of unions, parties, churches, and national as well as international NGOs and support groups, and is described as 'one of the quintessential social movements of the 20th century' (Stokke and Ballard 2005, p. 80). The recent history of social movement mobilisation in post-apartheid South Africa can broadly be divided into two phases. First, the 'honeymoon period' under Nelson Mandela's presidential term in office (1994–1999) was characterised by social movement de-mobilisation and absorption of anti-apartheid activists into state institutions (Leonard 2011, p. 116). The second phase since 2000 has witnessed the emergence of novel social movements marked by the beginning of the Mbeki era, continued through the Zuma presidency (1999–2007), and the deepening of the neoliberal reforms (Bond 2012). Neoliberalisation continued under the Zuma presidency (2009–2018), as he replaced Mbeki using some more nationalistic overtones, but not changing economic policy. Cyril Ramaphosa, who assumed office from Zuma in February 2018, began his tenure by entrenching neoliberal policy rather than bringing it to an end (Bond 2018a). Not least, Ramaphosa became one of the richest people in South Africa as a shareholder and board member of many multinational companies investing in mining, retail, communication, fast food, and many more industries (Butler 2013).

The deepening of neo-liberalisation and the end of the Mandela era was marked by social movement re-mobilisation. However, this time, there was no single enemy, but mobilisation revolved around a plurality of grievances (Ballard *et al.* 2006). The highest proportion of reported protests remains labour-related followed by community protests (Runciman *et al.* 2016). While these categories are broad and based on police reporting, they speak to the continuing relevance of labour in organising dissent in South Africa. However, many service-delivery protests and environmental protests against polluting industries (Leonard 2018; Lodge and Mottiar 2015; Habib and Opoku-Mensah 2009) are staged. While the study of labour movements has received a lot of attention in South African scholarship (Botiveau 2017; Marinovich 2016; Buhlungu 2010), novel formations, especially environmental social movements, are relatively understudied.

Scholars observe a relative decline of support for the African National Congress (ANC) over the years (Friedman 2018; Maharaj *et al.* 2011). South Africa under democracy offers various challenges as well as opportunities for social movement mobilisation. From the perspective of marginalised South Africans, the ANC's promise to redeem the injustices of apartheid has been largely disappointing, offering ample opportunities to remind the party where it failed (Stokke and Ballard 2005, p. 80). On the other hand, the post-apartheid government tends to channel dissent to deliberative fora. Participatory democratic governance inviting civil society organisation to participate in decision-making was strengthened in post-apartheid South Africa (Leonard 2014; Ballard 2007). Active participation of the citizenry was an express rallying point of the ANC, with grassroots organisations putting pressure on the newly formed government to stay true to the mass civic appeal that brought down apartheid (Friedman 2006, p. 1). From this vantage point, a tendency to capture and manufacture civil society actors has been observed. Leonard (2014, p. 381) shows that the 'post-apartheid state is financially supporting those types of individuals and organisations who are retreating from robust criticism'. Funding and support mainly go to larger civic organisations willing to provide services to the government, sometimes becoming a 'mirror image of state policies' (Leonard 2014, p. 379).

Because of the corporatist character of the post-apartheid South African state, dissent was partially absorbed into the Tripartite Alliance made up of the ANC, the umbrella Congress of South African Trade Unions (COSATU), and the South African Communist Party (SACP). As a result, even in some of today's service-delivery protest hotspots, such as Durban's Cato Manor, Merebank, and Wentworth, the ANC still manages to achieve good polling results (Lodge and Mottiar 2015).

Some civic organisations shifted their discourse to become more rights-based, affirming the Constitution. However, the rights-based approach runs the risk of demobilising civil society and weakening the grassroots appeal of social movements (Bond 2014; Leonard 2011, pp. 124–125). The Constitution's deliberate participatory spaces favour a prescribed corporatist engagement that further entrenches the grassroots' marginality. Friedman calls the existing participation mechanisms 'intrinsically hostile to effective participation of the poor' (Friedman 2006, p. 3). Moreover, commentators have observed that the ANC has shown a great aversion to oppositional extra-institutional politics, portraying activists as anti-democratic (Stokke and Ballard 2005, pp. 89–90; Habib and Opoku-Mensah 2009, p. 50).

Social movements challenging state-supported or state-sponsored practices are often characterised as anti-ANC and become criminalised. Oppositional grass-roots activists from the Landless People's Movement, Soweto Electricity Crisis Committee, the Shackdwellers' Movement, and anti-extractive groups have experienced repression and criminalisation (Ballard 2007, p. 20; Bond and Mottiar 2013; KaManzi and Bond 2015).

Post-apartheid social movement mobilisation is characterised by the absence of a common threat or unifying agenda. Jacklyn Cock, writing about the green movement in South Africa about a decade ago, observed the absence of a 'single, collective actor that constitutes the environmental movement in South Africa and no master "frame" of environmentalism encoded in any blueprint' (Cock 2007, p. 174). Yet, evidence from my research suggests that more sustained grassroots organisations are in the making. Part of the newness of emerging social movements is their interconnectedness within heterogeneous networks. The maintenance of these networks is not simple. Heterogeneity has also left social movements fragmented. Nevertheless, as I show in this book, socio-environmental struggles can align with the demands of different social groups. In light of accelerating fossil-fuel infrastructure expansion and climate crisis aggravation, movements of this kind are unlikely to disappear.

Structure of the Book

The remainder of the book is divided into two parts. The first part is about the theory and practice of resource extractivism. It looks at ways in which fossil-fuel extractivism is politically sustained in South Africa – both historically and today. The second part of the book is on social movements fighting fossil fuels – in South Africa, but also globally. Part I opens with Chapter 2, in which I present fossil-fuel frontiers as expanding general tendency underlying global capitalism involving extractivist practices. Its key characteristics are expansion and conflict. The deepening of resource frontiers undermines livelihoods and displaces residents and deteriorates the environment. The fossil-fuel frontier is a particularly destructive front that has its roots in industrialisation. However, while the fossil frontier creates environmental destruction, it has also contributed to economic development and prosperity in industrialised countries. The fossil-fuel economy enabled a consumer society that made life more comfortable worldwide for those lucky enough to have access to it. I highlight that the expansion of fossil fuels is organised by coercion and consent. Drawing on the Italian philosopher Antonio Gramsci, I stress that the expansion of fossil fuels is actively entrenched by dominant fractions in government, finance, and civil society.[1] I also show that any dominant social bloc needs to reach out to organise alliances and entrench its own interests possibly by using force.

In Chapter 3, I trace the emergence of fossil-fuel hegemony in South Africa from apartheid through to this day. After 1994, the ANC entrenched and expanded fossil-fuel extraction to fix developmental backlogs among historically disadvantaged South Africans rather than overcoming fossil fuel dependence. However, democratic elements in decision-making have been established to ensure formal participation in industrial projects such as mining. From a critical point of view, today's regulatory framework is biased towards favouring extraction.

While the first part of the book deals with the hegemonic structure of the fossil-fuel bloc, the second part analyses social movements fighting fossil fuels in South Africa and later offers a global outlook. Again, the second part will start with theoretical considerations. Chapter 4 starts with a literature review on 'spaces of social movement struggles' that is mainly concerned with the way space and place enable or constrain social movement mobilisation. The next section of the chapter links these insights on space to the notion of leadership. I present social movement leadership as a rather underdeveloped field of study. Leadership in social movements is central in giving direction and developing a sense of 'we'. Social movement leaders give political direction and frame the movement's demands. Chapter 4 closes with a discussion of the methods I used to analyse fossil-free movements in South Africa. During my fieldwork, I made use of a mix of qualitative methods including expert interviews, participant observations, and document analysis. In this part, I also reflect on my role as a white Western researcher in a postcolonial setting.

Chapter 5 discusses leadership practices and framing processes in the anti-coal movement in Fuleni bordering the oldest conservation area in Africa, the Hluhluwe–iMfolozi Park. In this chapter, I first trace the conflicted history of the area that is still known for its military battles between the Zulus and the British and later the eviction of local residents for the conservation park. In a second section of the chapter, I discuss framing processes of the Save our iMfolozi Wilderness campaign against coal mining bordering the conservation park. The leadership group of the movement changed the framing of the campaign several times in order to keep the campaign alive. This also happened when external shocks changed the movement's opportunity structure to make claims. I analyse the frame shifts referring to interviews with activists and written documents such as blog posts or press statements. I finally discuss why the Environmental Justice (EJ) framing that was adopted later in the campaign proves most successful in mobilising a broad movement campaign. EJ takes into view how polluting industries are often constructed in poorer neighbourhoods. This framing represents an experience beyond the fossil-fuel sector and offers potentials to connect the dots with other social movement campaigns.

In Chapter 6 on Frack Free South Africa (FFSA), I focus on anti-fracking movement tactics to stop the government-backed plans for hydraulic fracturing in the country. Movement tactics crucially give political direction to social movements. The choice of tactics will decide on the movement's success or failure to mobilise. FFSA used tactics to increase numbers as well as showing its worthiness to the outside. I discuss blocking and delaying, educating, connecting, and prefiguring as the four main tactics employed by FFSA. The analysis shows that many of the movement's resources are spent on information sharing, as fracking is not a very well-known technology in the South African context. Especially, declaring everyday places 'frack-free' is a low-level tactic to fight back potential intrusions of the fossil-fuel industry. Prefigurative practices are low-risk everyday actions portraying that a life without hydraulic fracturing is possible and desirable. In the conclusion and discussion of Chapter 6, I look at the potentials for concerted claim-making in between the two campaigns. I highlight that both SAVE and FFSA converge around the common goal of fighting fundamental changes in land-use practices of the fossil-fuel industry. Both movements share a commitment to nature conservation. However, it shows that they need to reach out to the wider public to broaden their appeal. The governing ANC that commands a majority in parliament stands firmly behind the 'fossil-fuel bloc'. Thus, the fossil-free movement needs to find allies to increase the pressure to de-carbonise the South African economy.

In Chapter 7, I offer a – perhaps surprising – analysis and comparison between Germany's and South Africa's fossil-free movements. Both countries are the number one producers of fossil fuels in Europe and Africa, respectively. In both countries, social movements thus stand against entrenched fossil-fuel interests. The German fossil-free movement managed to mobilise many activists forcing concessions from the national government. Finally, it will be of interest what are the potentials of different social movement frames regarding concerted claim-making between movements of the Global South and the Global North. I offer an analysis of the potentials of the Environmental Justice (EJ) frame comparing it to de-growth and Green New Deal (GND) frames. The book finally suggests that while fossil-free movements might not be the only hope for a post-fossil future, they certainly are the actors that address the issue of de-carbonisation most forcefully.

Note

1 I reject a demand-side approach that solely focuses on end consumers. According to the demand-side approach, consumers are responsible for the fossil-fuel economy, as they demand goods that are powered, produced, or delivered by fossil fuels. This demand-side view, however, overestimates the choice that consumers have, and it disregards the systemic nature of production and supply chains. Most consumers indeed do not have the choice to consume fossil-fuel-free goods and services, because supply chains are dripping of oil. Therefore, this study builds on the insight that the supply side of fossil fuel matters – i.e. the organisation of the energy and transport systems. Today's fossil-free movements are focusing increasingly on the supply side.

References

Alexander, P., *et al.*, 2012. *Marikana: a view from the mountain and a case to answer*. Johannesburg: Jacana Media.

Ballard, R., 2007. Participation, democracy and social movements. *Critical Dialogue*, 3 (2), 17–21.

Ballard, R., Habib, A. and Valodia, I., eds., 2006. *Voices of protest: social movements in post-apartheid South Africa*. Pietermaritzburg, South Africa: University of KwaZulu-Natal Press.

Bebbington, A., *et al.*, 2008. Contention and ambiguity: mining and the possibilities of development. *Development and Change*, 39 (6), 887–914. doi:10.1111/j.1467-7660.2008.00517.x

Blanco G., *et al.*, 2014. Drivers, trends and mitigation. *In:* O. Edenhofer, ed., *Climate change 2014: mitigation of climate change. contribution of working group III to the fifth assessment report of the intergovernmental panel on climate change*. Cambridge and New York: Cambridge University Press.

Bond, P., 2008. Social movements and corporate social responsibility in South Africa. *Development and Change*, 39 (6), 1037–1052. doi:10.1111/j.1467-7660.2008.00528.x

Bond, P., 2012. South African people power since the mid-1980s: two steps forward, one back. *Third World Quarterly*, 33 (2), 243–264. doi:10.1080/01436597.2012/020243-22

Bond, P., 2014. Constitutionalism as a barrier to the resolution of widespread community rebellions in South Africa. *Politikon*, 41 (3), 461–482. doi:10.1080/02589346.2014. 975931

Bond, P., 2018a, February 23. Cyril Ramaphosa relaunches neo-liberalism. *Pambazuka News*. Available from: www.pambazuka.org/economics/cyril-ramaphosa-relaunches-neo-liberalism

Bond, P., 2018b. Ecological-economic narratives for resisting extractive industries in Africa. *In*: P. Cooney and W.S. Freslon, eds., *Environmental impacts of transnational corporations in the global South*. Bingley: Emerald, 73–110.

Bond, P. and Mottiar, S., 2013. Movements, protests and a massacre in South Africa. *Journal of Contemporary African Studies*, 31 (2), 283–302. doi:10.1080/02589001.2013.78 9727.

Botiveau, R., 2017. *Organise or die? Democracy and leadership in South Africa's national union of mineworkers*. Johannesburg: Wits University Press.

Buhlungu, S., 2010. *A paradox of victory: COSATU and the democratic transformation in South Africa*. Scottsville, South Africa: University of Natal Press.

Burton, J. and Winkler, H., 2014. *South Africa's planned coal infrastructure expansion: drivers, dynamics and impacts on greenhouse gas emissions*. Cape Town: Energy Research Centre, University of Cape Town.

Butler, A., 2013. *Cyril Ramaphosa*. Johannesburg: Jacana.

Climate Action Tracker, 2018. Available from: https://climateactiontracker.org/countries/south-africa/pledges-and-targets

Cock, J., 2007. *The war against ourselves: nature, power and justice*. Johannesburg: Witwatersrand University Press.

Coryat, D. and Picq, M., 2016. Ecuador's expanding extractive frontier. *NACLA Report on the Americas*, 48 (3), 280–284. doi:10.1080/10714839.2016.1228184

Davenport, J., 2013. *Digging deep: a history of mining in South Africa*. Johannesburg: Jonathan Ball.

De Graaf, T. and Sovacool, B.K., 2020. *Global energy politics*. Cambridge: Polity.

Department of Energy, 2016. *Integrated resource plan update*. Available from: www. energy.gov.za/IRP/2016/Draft-IRP-2016-Assumptions-Base-Case-and-Observations-Revision1.pdf

Di Muzio, T., 2015. *Carbon capitalism: power, social reproduction and world order*. London: Rowman and Littlefield.

Eberhard, A., 2011. *The future of South African coal: market, investment and policy challenges* (Working Paper No. 100). Available from: https://pesd.fsi.stanford.edu/publications/the_future_of_south_african_coal_market_investment_and_policy_challenges

Engels, B. and Dietz, K., eds., 2017. *Contested extractivism, society and the state: struggles over mining and land*. London: Palgrave Macmillan.

Fine, B., 2008. *The minerals-energy complex is dead: long live the MEC?* Paper presented at Amandla Colloquium, Cape Town.

Fine, B. and Rustomjee, Z., 1996. *The political economy of South Africa: from minerals-energy complex to industrialisation*. London: Hurst and Company

Forslund, D. and Etkind, R., 2013. *Coping with unsustainability. Policy gap 7: Lonmin 2003–2012*. Johannesburg: Benchmarks Foundation.

Friedman, S., 2006. *Participatory governance and citizen action in post-apartheid South Africa* (Discussion Paper No. 164). Available from: www.ilo.org/wcmsp5/groups/public/-dgreports/-inst/documents/publication/wcms_193613.pdf

Friedman, S., 2018. Containing the crisis: what the ANC conference did – and did not do. *Transformation: Critical Perspectives on Southern Africa*, 96 (1), 95–101. doi:10.1353/trn.2018.0004

Goodman, J., *et al.*, 2020. *Beyond the coal rush: a turning point for global energy and climate policy*. Cambridge: Cambridge University Press.

Gramsci, A., 1991. *Selections from the prison notebooks*. London: Lawrence and Wishart.

Griffin, P., 2017. *CDP carbon majors report 2017*. The Carbon Majors Database. Available from: https://b8f65cb373b1b7b15feb-c70d8ead6ced550b4d987d7c03fcdd1d.ssl.cf3.rackcdn.com/cms/reports/documents/000/002/327/original/Carbon-Majors-Report-2017.pdf?1499691240

The Guardian Datablog, 2016. *World carbon dioxide emissions data by country: China speeds ahead of the rest*. Available from: https://www.theguardian.com/news/datablog/2011/jan/31/world-carbon-dioxide-emissions-country-data-co2

Habib, A. and Opoku-Mensah, P., 2009. Speaking to global debates through a national and continental lens: South African and African social movements in comparative perspective. *In*: S. Ellis and I. Van Kessel, eds., *Movers and shakers: social movements in Africa*. Leiden: Brill, 44–62.

Hallowes, D., 2011. *Toxic futures: South Africa in the crisis of energy, environment and capital*. Scottsville, South Africa: University of KwaZulu-Natal Press.

Hallowes, D., 2015. *Climate and energy: the elite trips out*. The Groundwork Report 2015. Available from: www.groundwork.org.za/reports/gW%20Report%202015.pdf

KaManzi, F. and Bond, P., 2015. Women from KwaZulu-Natal's mining war zone stand their ground against big coal (Report No. 23). *In:* L. Temper and T. Gilbertson, eds., *Refocusing resistance to climate justice: COPing in, COPing out and beyond Paris*. Available from: www.ejolt.org/2015/09/refocusing-resistance-climate-justice-coping-coping-beyond-paris

Klein, N., 2015. *This changes everything*. London: Penguin.

Lang, J., 1995. *Power base: coal mining in the life of South Africa*. Johannesburg: Jonathan Ball Publisher.

Larmer, M., 2010. Social movement struggles in Africa. *Review of African Political Economy*, 37 (125), 251–262. doi:10.1080/03056244.2010.51

Larmer, M. and Laterza, V., 2017. Contested wealth: social and political mobilisation in extractive communities in Africa. *The Extractive Industries and Society*, 4 (4), 701–706. doi:10.1016/j.exis.2017.11.001

Leonard, L., 2011. Civil society leadership and industrial risks: environmental justice in Durban, South Africa. *Journal of Asian and African Studies*, 46 (2), 113–129. doi:10.1177/0021909610391049

Leonard, L., 2014. Characterising civil society and its challenges in post-apartheid South Africa. *Social Dynamics*, 40 (2), 371–391. doi:10.1080/02533952.2014.936725

Leonard, L., 2016. State governance, participation and mining development: lessons learned from Dullstroom, Mpumalanga. *Politikon*, 44 (2), 327–345. doi:10.1080/02589346.2016.1245526

Leonard, L., 2018. Bridging social and environmental risks: the potential for an emerging environmental justice framework in South Africa. *Journal of Contemporary African Studies*, 36 (1), 23–38. doi:10.1080/02589001.2017.1412582

Leonard, L. and Pelling, M., 2010. Civil society response to industrial contamination of groundwater in Durban, South Africa. *Environment and Urbanization*, 22 (2), 579–595. doi:10.1177/0956247810380181

Lodge, T. and Mottiar, S., 2015. Protest in South Africa: motives and meanings. *Democratization*, 23 (5), 819–837. doi:10.1080/13510347.2015.1030397

Maharaj, B., Desai, A. and Bond, P., 2011. *Zuma's own goal: Losing South Africa's "war on poverty"*. Trenton, NJ: Africa World Press.

Malm, A., 2016. *Fossil capital: the rise of steam power and the roots of global warming*. London: Verso.

Marinovich, G., 2016. *Murder at small Koppie: the real story of South Africa's Marikana massacre*. London: Penguin.

Mining Africa, 2017. *Coal mining in Africa*. Available from: www.miningafrica.net/natural-resources-africa/coal-mining-in-africa/

Mitchell, T., 2013. *Carbon democracy*. London: Verso.

Oil Change International and Friends of the Earth United States, 2020. *Still digging: G20 governments continue to finance the climate crisis*. Washington. Available from: http://priceofoil.org/content/uploads/2020/05/G20-Still-Digging.pdf

Okonta, I. and Oronto, D., 2003. *Where vultures feast: shell, human rights, and oil in the Niger Delta*. London: Verso.

Oreskes, N. and Conway, E.M., 2012. *Merchants of doubt: how a handful of scientists obscured the truth on issues from tobacco smoke to global warming*. London: Bloomsbury Publishing.

Petras, J. and Veltmeyer, H., 2014. *The new extractivism: a post-neoliberal development model or imperialism of the twenty-first century?* London: Zed Books.

Petroleum Agency South Africa, n.d. *Exploration activities map*. Available from: www.petroleumagencysa.com/index.php/maps

Piggot, G., 2018. The influence of social movements on policies that constrain fossil fuel supply. *Climate Policy*, 18 (7), 942–954. doi:10.1080/14693062.2017.1394255

Pirani, S., 2018. *Burning up: a global history of fossil fuel consumption*. London: Pluto Press.

Runciman, C., *et al.*, 2016. *Counting police-recorded protests: based on South African police service data*. Johannesburg: Research Chair in Social Change, University of Johannesburg.

Sander, H., 2017. Ende Gelände: Anti-Kohle-Proteste in Deutschland. *Forschungsjournal Soziale Bewegungen*, 30 (1), 26–35. doi:10.1515/fjsb-2017-0004

Savides, M., 2017, June 19. *What the frack? Outrage over shale gas project near national park*. Available from: https://www.timeslive.co.za/news/south-africa/2017-06-19-what-the-frack-outrage-over-shale-gas-project-near-national-park/

Singer, M., 2011. Towards 'a different kind of beauty': responses to coal-based pollution in the Witbank coalfield between 1903 and 1948. *Journal of Southern African Studies*, 37 (2), 281–296. doi:10.1080/03057070.2011.579441

Snow, D.A. and Trom, D., 2002. The case study and the study of social movements. *In*: B. Klandermans and S. Staggenborg, eds., *Methods of social movement research*. Minneapolis: University of Minnesota Press, 146–172.

South African Market Insights, 2017, October 24. South Africa's coal exports: where is it going? *South African Market Insight*. Available from: www.southafricanmi.com/blog-24oct2017.html

Staggenborg, S., 2020. *Grassroots environmentalism*. Cambridge: University of Cambridge Press.

Stokke, K. and Ballard, R., 2005. Social movements in post-apartheid South Africa: an introduction. *In*: P. Jones, ed., *Democratising development: the politics of socio-economic rights in South Africa*. Leiden, the Netherlands: Koninklijke Brill, 77–100.

Temper, L., *et al.*, 2020. Movements shaping climate futures: a systematic mapping of protests against fossil fuel and low-carbon energy projects. *Environmental Research Letters*, 15 (12), 1–23. doi:10.1088/1748-9326/abc197

UNFCCC, 2016. *South Africa's intended nationally determined contribution (INDC)*. Available from: https://www4.unfccc.int/sites/ndcstaging/PublishedDocuments/South%20Africa%20First/South%20Africa.pdf

Van Wyk, D., 2013. Globalization and the minerals industry: a South African case study. *In*: L. Masters and E. Kisiangani, eds., *Natural resources governance in Southern Africa*. Pretoria: Africa Institute of South Africa, 3–34.

World Bank, 2018. *Electricity production from coal sources (% of total)*. Available from: https://data.worldbank.org/indicator/EG.ELC.COAL.ZS?locations=ZA

Yeld, J., 2018, July 24. Coal mine opponents targeted on social media. *GroundUp*. Available from: www.groundup.org.za/article/death-threat-against-environmental-activist/

Part I

Theory and Practice of Resource Extractivism

2 Resource Frontiers and Hegemony

This chapter traces the drivers of resource extraction that take place on an ever-larger scale globally and the power structures enabling global extractivism. Natural resources only 'become' resources when humans decide to use them for production and consumption (Vorbrugg 2019). Throughout human history, natural resources from biomass to fossil fuels have always been used. However, the intensity and scale of resource extraction determine the impact on the natural environment and therefore also human livelihoods. An examination of resource uses should bear in mind the carrying capacity of the planet. Johan Rockström and co-authors define nine threshold levels for a safe operating space for humanity (Rockström 2009). Three out of nine threshold levels were transgressed when writing this book (biodiversity loss, nitrogen cycle, and climate change) (Steffen *et al.* 2015). The planetary boundaries of land use and other parameters are increasingly at risk. From a critical economy perspective, overstepping planetary boundaries is intimately linked to the global growth paradigm (Hickel 2019).

Political power relies on access to the critical natural resource for industry, transport, agriculture, and households, among other sectors. Access and appropriation of resources are central to societies' rise or collapse (Diamond 2011). Empires, kingdoms, and democracies flourished when they had sufficient access to resources (Diamond 2011; Bunker and Ciccantell 2005).

These insights will be unpacked by thinking through the relationship between resources and political power in this chapter. The question of fossil fuels as resources for political power will be of particular interest. To encapsulate the expansion of resource extraction, I will introduce the notion of the fossil-fuel frontier understood as a physical boundary and a political project. Frontier expansion is usually accomplished using a standardised mapping approach resulting in surveying, dispossession, and treaties to normalise frontier expansion.

In the second section, I discuss Antonio Gramsci's notion of hegemony as a product of coercive state power and persuasion that includes civil society. Gramsci's notion of the 'integral state' comes to grips with the complex alliance between state and civil society. In 'organic crises', this alliance can lose its grip. Building on post-Gramscian scholars such as Laclau and Mouffe (2001), I show how social movements split the social into antagonistic camps by articulating differences between the 'us' and 'them'. Hegemonic social forces will try to strengthen their respective social position in response.

DOI: 10.4324/9781003110835-3

Resource Frontiers

In general terms, frontiers demarcate the limits of something; for example, one country from another, or the edge of a land beyond which the 'country is wild and unknown' as Frederick Turner wrote at the end of the 19th century with respect to the significance of the American frontier (Turner 1958). Like country borders, resource frontiers are the outcome of a historical process presupposing measurement, mapping, conquest, and treaties. Resource frontiers also divide the known from the unknown, proven resources from unmapped spaces. Therefore, knowledge of geography plays a significant role in frontier expansion. Woodworth highlights the value of the resource frontier concept 'for critical interpretations of the shifting geographies of development agendas' (2017, p. 136).

Capitalism presupposes the active expansion of profitable accumulation strategies, especially natural resource industries, as the basis for today's fossil-fuel-based civilisation (Baglioni and Campling 2017; Di Muzio 2015; Moore 2000). As Heim puts it: 'The system's ongoing growth depends on incessantly redrawing its boundaries' (1996, p. 687). Frontiers of capitalist expansion enclose lands for accumulation and commodification of ever more aspects of everyday life.[1] New resource frontiers introduce 'brutal geographies' (Brooks 2005) into frontier spaces, profoundly changing social, environmental, and economic relations. The concrete frontiers are created by market prices of commodities, global demand, and investment flows into a specific accumulation strategy. New resource frontiers are expanded faster in deeper seas and even in outer space (Shaw 2013). Fossil fuels, water, and rare earth elements are among the fastest accelerating frontiers of our times (Ruz 2011).

Like frontier expansion in the 19th century, resource frontiers do have violent elements to them. Hegemonic actors tend to gloss over the concrete repercussions of frontier expansion such as environmental degradation, air and water pollution, and land grabbing. Instead, they emphasise procedures, laws, employment figures, GDP growth, and deadlines for objections. Expanding the frontier is advanced as a 'common sense' subject only to legal and technocratic requirements. This standardised approach is common to high-modern resource projects (Schilling-Vacaflor *et al.* 2018; Scott 1998). Scott makes the case against what he calls 'an imperial or hegemonic planning mentality that excludes the necessary role of local knowledge and know-how' (Scott 1998, p. 6).

The process of frontier expansion can take three possible forms. First, frontier expansion can take the form of 'accumulation by dispossession', understood as the expansion of frontiers without adequate compensation or consent as seen in different mining regions around the developing world (Andrews 2018; Allen 2017; Perreault 2013; Bebbington *et al.* 2008). Second, and less acknowledged in the literature, frontier expansion can take the form of struggles between different accumulation strategies. Actors may propose expanding two or more mutually exclusive forms of frontiers (i.e. coal mining and conservation). Third, consent is possible by accommodating local demands and engagement of all relevant actors (Yakovleva and Vazquez-Brust 2012; Cragg and Greenbaum 2002; Weber-Fahr

2002; Humphreys 2000; Hilson and Murck 2000). The first two forms of resource frontier expansion take the form of struggle, while the third amounts to consent to resource expansion.

Frontier struggles create a liminal space 'separating the usual from the unusual and the possible from the impossible' (Büscher 2013, p. 9). The possibilities of the frontier space are fought over and negotiated between different actors. Van Teijlingen and Hogenboom (2014, p. 7), for example, describe mining conflicts as

> involving a variety of stakeholders who operate at different scales and have diverse notions of the relationship between society and nature, often result-ing in different and conflicting spatial practices and territorial claims. . . . these include tangible aspects such as the use of land and water, as well as non-tangible aspects such as knowledge, histories, cultures, value systems and ultimately notions of what nature, development and territoriality mean.

Scholars of frontier struggles need to engage closely with the respective actor constellation on the ground. As Büscher (2013, p. 10) rightly put it: 'opening new frontiers is a profoundly political act', and therefore understanding power relations in a given frontier space is vital. Scholars observe that hegemonic actors try to normalise resource extraction by 'depoliticising' bureaucratic practices (Schilling-Vacaflor *et al.* 2018; Friedman 2006).

The resource frontier approach has some overlap with transition approaches but also differs in some crucial respects. Transition approaches focus on resource regimes like the energy sector and suggest that intra-institutional processes might trigger longer-term policy changes (Baker *et al.* 2014; Newell and Mulvaney 2013; Meadowcroft 2009; Winkler and Marquand 2009). More particularly, old energy regimes like the minerals-energy complex (MEC) in South Africa might be poten-tially superseded by niche innovations, culminating in changes in the dominant design influencing policy, technology, culture, and science in the long term (Baker *et al.* 2014, p. 5). Both the transition approach and the frontier approach acknowl-edge the inherently political dimension of energy policy and the influence of pow-erful hegemonic actors. Both approaches also share that energy transitions might happen through institutional innovation. However, three major differences com-pare the transition approach to the notion of the resource frontier as proposed in this book. First, the literature on renewable energy transitions looks at how renew-able 'niche' developments can be beneficial for policymakers and end-users alike (Baker 2015; Baker *et al.* 2014). The fossil-fuel frontier approach explores how the fossil-fuel economy maintains itself and understands how civil society actors try to enlarge their counter-hegemonic 'chain of equivalence' by contesting the dominant social forces. The frontier approach suggests that, in principle, there can be accom-modation in the energy regime, but it also accounts for conflict and hegemonic struggles (Watts 2018; Büscher 2013). Second, both approaches also differ in the concrete actors they identify as resisting the incumbent regime. While the energy transition literature focuses on renewable energy producers enlarging their role in the energy market, the frontier approach looks primarily at hegemonic actors in the private sector, the state, and the role of social movements in challenging existing energy pathways. The emphasis on social movements as change agents is justified given the ossified institutional structures marginalising non-fossil-fuelled energy procurement. Despite the fact that South Africa has seen system-changing social movements in the past, the energy transition approach does not identify them as potential sources of change. Instead, the transition approach looks at the roles of existing institutions in government and the private sector (Baker 2015; Baker *et al.* 2014). Third, the methodology of the approaches is slightly different. The focus of the transition literature is on possible policy pathway changes and primarily looks at policy documents. In contrast, the frontier approach presented here adds ethnographical insights and focuses on concrete places of possible extraction.

I will argue that frontier struggles take different forms in different places, often overlooked by other scholars describing resource expansion as a one-way process advanced by governments and companies. To understand the resource frontier as the outcome and process of hegemonic space making, I will revisit Gramsci's rich theoretical vocabulary in the following.

Organising Hegemony

Antonio Gramsci's political thought holds valuable insights into thinking about the formation of state power and its maintenance (Glassman 2013; Karriem 2009; Gramsci 1991). Unlike some more orthodox renditions of Marxism, Gramsci offers a grammar to dynamically study the formation of political alliances, paying special attention to spatial differentiation. Writing from a fascist Italian prison between 1929 and 1935, Gramsci argues that coercive state power exercised by force is not enough to uphold hegemony. The coercive force needs to be complemented by consent. According to Gramsci, the site of persuasion is civil society. Hegemony cannot be exercised on 'unwilling' subjects but is the result of a concrete alignment between the state and civil society (Thomas 2009, p. 163), which Gramsci illustrates with his analyses of the Italian Risorgimento and the French Revolution (Gramsci 1991, pp. 53, 77–79). Gramsci observed that in Italy in the 1920s, hegemony was exercised through the exploitation of the working classes and through cultural forms such as the bourgeois education system and the press 'unconsciously' working on the oppressed masses' minds (Gramsci 1926, p. 3). He was, thus, very attuned to the interlinking of actors struggling for hegemony at different scales. Crucially, Gramscian scholars point out that hegemony works on different scales and that studies of hegemony need to examine the interaction of actors on different geographical levels (Kontinen and Millstein 2016; Karriem 2009).

Gramsci defines the 'historical bloc' as 'the complex, contradictory and discordant ensemble of the superstructures . . . the reflection of the *ensemble* of the social relations of production' (Gramsci 1991, p. 366). 'Historical blocs' are characterised by a dialectical tension between the ideological superstructure and material base. They are both constrained by history and, at the same time, subject to contestation to its hegemonic composition (Gramsci 1991, p. 366).

Saull (2012, p. 336) argues that the 'pattern of hegemony is founded on the relationship and the contradictions that develop between the regime of material accumulation and the social, political, ideational, and institutional fabric of the historical bloc'. Hegemony has two elements: the 'structural/material' and the 'conjunctural/agential' (Saull 2012). Gramsci offers a rich vocabulary, pointing out mechanisms through which hegemony is constituted, sustained, and challenged.

Gramsci calls the formation of hegemonic alliances between the state and civil society 'integral state'. According to Mouffe, Gramsci made at least two important additions to the study of hegemony. First, Gramsci privileges 'ideological superstructure over the economic structure', and second, points to the 'primacy of civil society (consensus) over political society (force)' (Mouffe 1979, p. 3). In the following statement, the two elements are thought of together as the:

> product of leadership . . . a consequence of individual and human acts, the Gramscian reading of this concept draws our attention to both its contestability and the impossibility of reducing it to a preponderance of material resources.
>
> (Germain and Kenny 1998, p. 6)

Purposeful and persuasive leadership is a key in formulating and maintaining hegemony. A hegemonic bloc fosters a common worldview while serving narrow class interests. This worldview must be reasonably open and leave space for a degree of pluralism (Mouffe 1979, p. 15). Crucial to the fostering of a durable political culture is the work of 'organic intellectuals'. Rather than being confined to a particularly learned elite, intellectualism is a form of 'philosophy of praxis'. Gramsci points out that every social formation has intellectuals. A standard form of consciousness to arrive at a 'higher level of one's own conception of reality' is a prerequisite for the gradual formulation of a more encompassing worldview (Gramsci 1991, p. 333). Organisers and leaders take the role of disseminating ideas and embodying the hegemonic bloc in political praxis. 'Traditional intellectuals', on the other hand, have formal qualifications but do not see themselves as part of a social formation, although they play crucial roles in sustaining social orders (Gramsci 1991, p. 7).

Gramsci maintains that at crucial political junctures, the hegemonic bloc loses its appeal, meaning 'that the great masses have become detached from their traditional ideologies, and no longer believe what they used to believe previously' (Gramsci 1991, p. 275). In times of hegemonic crises, the ruling classes often have nothing left but the use of force; and as Gramsci famously proclaims: 'the old is dying and the new cannot be born; in this interregnum, a great variety of morbid symptoms appear' (1991, p. 276). Crises afford opportunities to refashion new societal relations. For Gramsci, however, the outcomes of crises are contingent. For any successful political movement, Gramsci argues that the right choice of strategies is pivotal. While the balance of forces allowed for a frontal attack on the state in czarist Russia in 1917, a 'war of position' would be necessary for more advanced bourgeois societies with consolidated civil societies (Roger 1991, p. 75; Gramsci 1991, p. 237). A 'war of position' takes place on the ideological plane. It is the consciousness-shaping process that claims positions 'apart' from the hegemonic bloc. Inspired by 'an instinctive feeling of independence, and which progresses to the level of the real possession of a single and coherent conception of the world' (Gramsci 1991, p. 333), counter-hegemonic forces challenge the *status quo*. To displace taken-for-granted assumptions, the counter-hegemonic bloc needs to challenge the 'common sense' prevalent in society and overcome it by crafting a conscious 'good sense'. In contemporary societies, market relations, for example, are deeply entrenched in forms of everyday life and need to be 'refracted through actual discursive practices, understandings, and behaviours' (Harvey 2006, p. 84). Creating a 'good sense' for Gramsci is only possible by engaging in a 'philosophy of praxis': a way of thinking and acting differently (Gramsci 1991, pp. 323–343). Gramsci's philosophy remains relevant today, as he answers crucial questions regarding how political formations are maintained and might be resisted in the future. Scholars inspired by Gramsci spell out ways to think through hegemony, including writers in the political ecology tradition, focusing on the concrete ways in which the environment becomes the object of hegemonic struggles (Perkins 2011; Mann 2009).

In *Hegemony and Socialist Strategy* and later publications, Laclau and Mouffe (2001) 'completed' Gramsci's project (Legett 2013, p. 301). While Laclau and Mouffe's contributions to the study of hegemony admittedly remain 'notoriously abstract and removed from the realities of social movements and their concrete resistance struggles against hegemonic orders' (Livramento Dellagnelo *et al.* 2014, p. 144), important conceptual lessons can be learnt from their work, especially when it comes to questions of alliance-building and political strategy. According to Laclau and Mouffe (2001), any given societal order is never fully complete. Instead, social orders can only ever be partial and, thus, open to challenges. They propose to study hegemony discursively as systems of meaning open to re-articulation. The totality of articulatory practices they call discourse embraces 'all social practices and relations' (Howarth 2000, p. 101). Hence, Laclau and Mouffe (2001) are committed to a radically non-foundational understanding of political practices. The authors emphasise the 'non-closure of the social' (Norval 1997, p. 53), meaning that any given social formation is never fully hegemonised. Norval, building on Laclau and Mouffe's insights, explains that political frontiers are particularly pronounced in spaces in which 'centralised forms of oppression endow popular struggles with clearly defined enemies' (Norval 1997, p. 62). The drawing of political frontiers becomes more pronounced in spaces where a clear distinction between 'us' and 'them' can be identified.

Central to their argument is that hegemonic forces are the outcome of articulatory practices assembling 'chains of equivalence'. 'Chains of equivalence' are structured around nodal points also referred to as 'empty signifiers' (Laclau 2006;

Laclau and Mouffe 2001). They emphasise that political frontiers are the outcome of the active creation of concrete discursively constructed alliances. In Mouffe and Laclau's thinking, political struggles do not follow essentialist class categories underpinned by objective, material realities that inevitably pit the bourgeoisie against the proletariat. The failure of a hegemonic system to satisfy requests might end up in 'an accumulation of unfulfilled demands and an increasing inability of the institutional system to absorb them in a *differential* way (each in isolation from the others), and an *equivalential* relation is established between them' (Laclau 2006, p. 73). With growing estrangement from the institutional logic of demands, there is an opportunity to forge equivalential links between counter-hegemonic forces. Different elements within the system of meaning are rendered equal by chains of equivalences, effectively splitting the discursive system into 'two opposed camps' (Howarth and Stavrakakis 2000, p. 1).

Laclau and Mouffe see neither a necessary essence nor determination (e.g. class struggle) in the drawing of political frontiers. These frontiers can be drawn across multiple identities, classes, and interests. Social formations and alliances for Laclau and Mouffe are never predetermined. The formation of hegemonic alliances is structured around emptiness. While a nodal point eventually comes to stand for the system as a whole, there is no necessary connection between the signified system and signifying political actors.

Inspired by Gramsci's reading of Marx, Mouffe and Laclau make valuable contributions by proposing how political subjects form alliances discursively. Chains of equivalence encompass heterogeneous demands with no necessary internal relationship. Enlarging the chain is necessary to formulate an antagonistic social struggle and is also dangerous, as individual demands become ever less important. This tension between the logic of equivalence and their difference is a key insight in understanding the formation of heterogeneous social alliances. The concept of the empty signifier points at how social alliances group around particular moments within a system of meaning. In his later work, Laclau becomes particularly interested in how emptiness can be embodied by particular leaders (Laclau 2007).

However, by discussing struggles solely at an abstract discursive level, Mouffe and Laclau neglect the importance of material underpinnings of social struggles (Livramento Dellagnelo *et al*. 2014, p. 144). Mouffe and Laclau applaud Gramsci for freeing Marxism from 'epiphenomenalism (viewing politics as secondary to the economic infrastructure) and economic reductionism (seeing political activity as determined by given class relations)' towards a 'more open-ended Marxism' (Legett 2013, p. 301). They take the project of non-foundationalism to the extreme end of post-structuralism. In other words, by 'radicalising' Gramsci's residual essentialist categories, they 'free' hegemony from any deterministic 'baggage'. Gramsci still insisted on 'fundamental classes' determined by the mode of production in a given society, thus stressing that material conditions shape social relations. By insisting that struggles have no necessary material base, Laclau and Mouffe (2001) take their post-foundational project one step too far. Žižek (2000, p. 97) criticises the underlying postmodernism of Laclau's interpretation of Gramsci for displacing class struggle as just another struggle next to anti-sexist struggles or gay rights protests, for example. Veltmeyer (2000, p. 510) makes the same point, juxtaposing Laclau's thinking with Gramsci's writings:

> for Gramsci no matter how complex the contingent mediations involved at the level of theory and practice (and culture and ideology) the concept of hegemony is fundamentally tied to the existence of classes formed in the sphere of economic production, as well as a politically defined class struggle.

Indeed, the few empirical examples that Laclau gives in his writing suggest an underpinning material kernel at the centre of political mobilisation. Take the following example:

> Think of a large mass of agrarian migrants who settle in the shantytowns on the outskirts of a developing industrial city. Problems of housing arise, and the group of people affected by them request some kind of solution from the authorities. Here we have a *demand* which initially is perhaps only a request. If the demand is satisfied, that is the end of the matter; but if it is not, people start to perceive that their neighbours have other, equally unsatisfied demands – problems with water, health, schooling and so on.
>
> (Laclau 2007, p. 73)

The earlier argument takes the realm of political economy and an understanding that basic needs shape political struggles seriously. I suggest that it is helpful to retain Gramsci's understanding that accumulation strategies have a bearing on articulating political grievances and demands. Veltmeyer (2000) observes that while Gramsci's understanding of hegemony is grounded in historical materialism, Laclau and Mouffe have abandoned Marxism, supposedly retaining its 'best fragments' (Veltmeyer 2000).

It is helpful to look at 'chains of equivalence' to understand that counterhegemonic formations are the outcome of articulatory practices. Laclau and Mouffe (2001) offer important insights in analysing how social struggles work internally. Counter-hegemonic formations are the *outcome* of the articulation of different grievances and demands along a chain of equivalence. Through a reading of Gramsci, they show how these chains are indeed assembled as an outcome of organic leadership.

Adding to this point, I wish to foreground Gramsci's appreciation of social struggles, giving special attention to geography. While Gramsci's writings speak of the need to appreciate hegemony in concrete political junctures, Mouffe and Laclau's inquiries often lack precisely that. Places and their meaning are historically marked, as Marx famously said in his *Eighteenth Brumaire of Bonaparte*. It is also helpful to remember Marx's insistence on the role of historical circumstances that have bearings on today's political junctures, stressing the link between past and present thinking.

> Men make their own history, but they do not make it as they please; they do not make it under self-selected circumstances, but under circumstances exist-ing already, given and transmitted from the past. The tradition of all dead generations weighs like a nightmare on the brains of the living.
>
> (Marx 1852)

The struggle over hegemony takes form in concrete time and space. According to Gramsci, insurgent strategists need to take the political conjuncture they find themselves in seriously. Gramsci was highly attuned to the interaction between differently organised geographical spaces and political formations, as his writ-ing on northern and southern Italy show (Gramsci 1926). Gramsci thinks, on the one hand, that unevenness creates differently structured political subjectivi-ties, whereas, on the other hand, he thinks of space rather dynamically. Personal exchanges across differently structured spaces foster alliances (Featherstone 2013, p. 68; Gramsci 1926). The challenge for Gramsci outlined in his considerations on the 'Southern Question' (1926) becomes how an alliance between the impover-ished peasantry and the city could be fostered. When the southern peasantry was subservient to the northern bourgeoisie, Gramsci called the relation between the two parts of the country 'colonial'.

Gramsci's conceptual framework can also be employed to study hegemony from a comparative perspective. This view renders visible the concrete ways hegemony takes shape in different places rather than assuming that hegemony is constant or unspecific to geographic specificities. Avcı (2015), for example, compares mining struggles in Ecuador with mining struggles in Turkey, looking at how the respective political geographies shape the affected peasants' experi-ence of mining. While the Turkish state exercises hegemony 'on the basis of the ideal of modernisation through economic development . . . ideological leader-ship, centralised and paternalistic governance, and material concessions that have improved the living standards of even the most impoverished classes' (Avcı 2015, p. 322), the Ecuadorian state does not command the same degree of economic and ideological dominance. However, since 2007, with Rafael Correa taking power in Ecuador, hegemony in the two states has become more alike. Correa linked developmental success closely to an extractivist agenda. Comparing two resistance movements at Mount Ida in Turkey and Intag in the Ecuadorian Andes, Avcı (2015) observes stark differences in the ways the movements are organising anti-mining resistance. Resistance at Mount Ida is driven by 'middle-class profes-sional organisations, amenity migrants, and local government most of which are from the main opposition party', effectively excluding local peasants from engag-ing meaningfully in this process (Avcı 2015, p. 321).

In contrast, opposition movements in Intag grew based on communities where national conservation groups traditionally employed militant techniques to combat. Later, these groups were joined by local government, indigenous groups, and international anti-mining networks (Avcı 2015, p. 318). Avcı shows that in the Ecuadorian case, the ambition to stop mining has a broader scope. She explains this by referring to 'close and democratic relationships' between a cross-section of different actors, 'their long-term co-operation beyond the problem of mining', and the leadership role that local peasant activists play (Avcı 2015, p. 323). This feature is not present in the movement at Mount Ida, where 'relations between the peasants, civil society, and local government actors have remained rather distant and hierarchical' (Avcı 2015, pp. 323–324).

Karriem (2009), in his analysis of the Brazilian Landless Rural Movement (MST), uses a Gramscian lens to stress the importance of the 'philosophy of praxis' in an agro-ecological counter-hegemonic formation. As a crucial reason for MST's mobilising success, Karriem identifies the scaling up of protest on different occasions, such as at transnational anti-World Trade Organization (WTO) rallies and strategic alliances with other peasant movements such as La Vía Campesina. Karriem highlights that these allies' interventions have forced multinational corporations such as Monsanto to backtrack from some of their experimental agro-industrial plantations (2009, p. 322). Tracing the MST rise from a local to a global counter-hegemonic actor, Karriem highlights that the insurgent campaigns cannot be seen simply as linear struggles towards emancipation. Rather,

> MST's praxis since its founding has been marked by conquests and defeats, offensive and defensive struggles, loss of lives and the conquest of land, which reflects a Gramscian war of position as a long, slow process of practical and ideological struggle for an alternative hegemony.
>
> (Karriem 2009, p. 324)

Therefore, the study of hegemony has to be sensitive to the ebbs and flows of a movement and must trace counter-hegemonic articulations over time. More recently, Gramsci's concepts find growing resonance with scholars analysing multi-scalar hegemonic formations in Africa (Brooks and Loftus 2016; Hart 2014). Broadly, these texts on Malawi and South Africa, respectively, highlight that contemporary elites change and rearrange their hegemonic position through 'passive revolution'. Using the term 'passive revolution', Gramsci refers to moments of great societal changes that leave the central tenets of property relations at a given political juncture intact (Gramsci 1991, p. 106). The rise of fascism in Italy is a case in point for Gramsci. He observes how the Italian bourgeoisie could forge alliances and quell popular demands by reinforcing 'the hegemonic system and the forces of the military and civil coercion at the disposal of the traditional ruling classes' (Gramsci 1991, p. 120).

Similarly, Brooks and Loftus (2016) argue that the accession of the president in Malawi in 2012 was necessary for the ruling classes to leave intact societal relations skewed towards the ruling elite. Hart (2014, p. 220) suggests that white capital was left untouched in South Africa during the transition period in the 1990s.

Hart adds that anyone studying post-colonial settings in African hegemony fostered by 'passive revolution' needs to be attentive to the 'ongoing reverberations of colonial histories and changing forms of imperialism; and to race, ethnicity, gender and sexuality' (Hart 2014, p. 224). Gramscian scholars working on Africa stress that the concepts employed by the Sardinian intellectual and political organiser are sensitive to the specificities of time and space and offer rich perspectives on different African political formations.

Note

1 The frontiers of capital are obviously not all material, but also take immaterial forms such as commodification of data by big tech companies (Beverungen *et al*. 2015), financialisation (Foster 2022), and even commodification of pollution as carbon trading (Bryant *et al*. 2015), for example.

References

Allen, M.G., 2017. Islands, extraction and violence: mining and the politics of scale in Island Melanesia. *Political Geography*, 57, 81–90. doi:10.1016/j.polgeo.2016.12.004

Andrews, N., 2018. Land versus livelihoods: community perspectives on dispossession and marginalization in Ghana's mining sector. *Resources Policy*, 58, 240–249. doi:10.1016/j.resourpol.2018.05.011

Avcı, D., 2017. Mining conflicts and transformative politics: a comparison of Intag (Ecuador) and Mount Ida (Turkey) environmental struggles. *Geoforum*, 84, 316–325. doi:10.1016/j.geoforum.2015.07.013

Baglioni, E. and Campling, L., 2017. Natural resource industries as global value chains: frontiers, fetishism, labour and the state. *Environment and Planning A: Economy and Space*, 49 (11), 2437–2456. doi:10.1177/0308518x17728517

Baker, L., 2015. Renewable energy in South Africa's minerals-energy complex: a 'low carbon' transition? *Review of African Political Economy*, 42 (144), 245–261. doi:10.1080/03056244.2014.953471

Baker, L., *et al*., 2014. The political economy of energy transitions: the case of South Africa. *New Political Economy*, 19 (6), 791–818. doi:10.1080/13563467.2013.849674

Bebbington, A., *et al*., 2008. Contention and ambiguity: mining and the possibilities of development. *Development and Change*, 39 (6), 887–914. doi:10.1111/j.1467-7660.2008.00517.x

Beverungen, A., Böhm, S. and Land, C., 2015. Free labour, social media, management: challenging Marxist Organization Studies. *Organization Studies*, 36 (4), 473–489. doi:10.1177/0170840614561568

Brooks, S., 2005. Images of 'Wild Africa': nature tourism and the (re)creation of Hluhluwe game reserve, 1930–1945. *Journal of Historical Geography*, 31 (2), 220–240. doi:10.1016/j.jhg.2004.12.020

Brooks, A. and Loftus, A., 2016. Africa's passive revolution: crisis in Malawi. *Transactions of the Institute of British Geographers*, 41 (3), 258–272. doi:10.1111/tran.12120

Bryant, G., Dabhi, S. and Böhm, S., 2015. 'Fixing' the climate crisis: capital, states, and carbon offsetting in India. *Environment and Planning A: Economy and Space*, 47 (10), 2047–2063. doi:10.1068/a130213p

Bunker, S.G. and Ciccantell, P.S., 2005. *Globalization and the race for resources*. Baltimore: The Johns Hopkins University Press.

Büscher, B., 2013. *Transforming the frontier: peace parks and the politics of neoliberal conservation in Southern Africa*. Cape Town: University of Cape Town.

Cragg, W. and Greenbaum, A., 2002. Reasoning about responsibilities: mining company managers on what stakeholders are owed. *Journal of Business Ethics*, 39, 319–335. doi:10.1023/A:1016523113429

Diamond, J., 2011. *Collapse: how societies choose to fail or survive*. London: Penguin.

Di Muzio, T., 2015. *Carbon capitalism: power, social reproduction and world order*. London: Rowman & Littlefield.

Featherstone, D., 2013. 'Gramsci in action': space, politics, and the making of solidarities. *In*: M. Ekers, G. Hart, S. Kipfer and A. Loftus, eds., *Gramsci, space, nature, politics*. Chichester, England: Wiley-Blackwell, 65–82.

Foster, J.B., 2022, March. The defense of nature: resisting the financialization of the earth. *Monthly Review*, 73 (11).

Friedman, S., 2006. *Participatory governance and citizen action in post-apartheid South Africa* (Discussion Paper No. 164). Available from: www.ilo.org/wcmsp5/groups/public/-dgreports/-inst/documents/publication/wcms_193613.pdf

Germain, R.D. and Kenny, M., 1998. Engaging Gramsci: international relations theory and the new Gramscians. *Review of International Studies*, 24 (1), 3–21. doi:10.1017/s0260210598000035

Glassman, J., 2013. Cracking hegemony: Gramsci and the dialectics of rebellion. *In*: M. Ekers, G. Hart, S. Kipfer and A. Loftus, eds., *Gramsci, space, nature, politics*. Chichester, England: Wiley-Blackwell, 241–257.

Gramsci, A., 1926. *Some aspects of the southern question*. Available from: https://cpbuse1.wpmucdn.com/blogs.uoregon.edu/dist/f/6855/files/2014/03/gramsci-southernquestion1926-2jf8c5x.pdf

Gramsci, A., 1991. *Selections from the prison notebooks*. London: Lawrence and Wishart.

Hart, G., 2014. *Rethinking the South African crisis: nationalism, populism, hegemony*. Athens, GA: University of Georgia Press.

Harvey, D., 2006. *Spaces of global capitalism: towards a theory of uneven geographical development*. Brooklyn: Verso.

Heim, C.E., 1996. Accumulation in advanced economies: spatial, technological, and social frontiers. *Cambridge Journal of Economics*, 20 (6), 687–714. doi:10.1093/oxfordjournals.cje.a013645

Hickel, J., 2019. Is it possible to achieve a good life for all within planetary boundaries? *Third World Quarterly*, 40 (1), 18–35. doi:10.1080/01436597.2018.1535895

Hilson, G. and Murck, B., 2000. Sustainable development in the mining industry: clarifying the corporate perspective. *Resources Policy*, 26 (4), 227–238. doi:10.1016/s0301-4207(00)00041-6

Howarth, D.R., 2000. *Discourse*. Buckingham: Open University Press.

Howarth, D.R. and Stavrakakis, Y., 2000. Introducing discourse theory and political analysis. *In*: D.R. Howarth, A.J. Norval and Y. Stavrakakis, eds., *Discourse theory and political analysis: identities, hegemonies, and social change*. Manchester: Manchester University Press, 1–23.

Humphreys, D., 2000. A business perspective on community relations in mining. *Resources Policy*, 26 (3), 127–131. doi:10.1016/s0301-4207(00)00024-6

Karriem, A., 2009. The rise and transformation of the Brazilian landless movement into a counter-hegemonic political actor: a Gramscian analysis. *Geoforum*, 40 (3), 316–325. doi:10.1016/j.geoforum.2008.10.005

Kontinen, T. and Millstein, M., 2016. Rethinking civil society in development: scales and situated hegemonies. *Forum for Development Studies*, 44 (1), 69–89. doi:10.1080/08039410.2016.1264994

Laclau, E., 2006. Why constructing a people is the main task of radical politics. *Critical Inquiry*, 32 (4), 646. doi:10.2307/3877130

Laclau, E., 2007. *On populist reason*. London: Verso.

Laclau, E. and Mouffe, C., 2001. *Hegemony and socialist strategy: towards a radical democratic politics*. London: Verso, 2nd ed.

Legett, W., 2013. Restoring society to post-structuralist politics: Mouffe, Gramsci and radical democracy. *Philosophy & Social Criticism*, 39 (3), 299–315. doi:10.1177/0191453712473080

Livramento Dellagnelo, E.H., Böhm, S. and Mendonça, P.M., 2014. Organizing resistance movements: contribution of the political discourse theory. *Revista de Administração de Empresas*, 54 (2), 141–153. doi:10.1590/s0034-759020140203

Mann, G., 2009. Should political ecology be Marxist? A case for Gramsci's historical materialism. *Geoforum*, 40, 335–344. doi:10.1016/j.geoforum.2008.12.004

Marx, K., 1852. *The eighteenth Brumaire of Louis Bonaparte*. Available from: www.marxists.org/archive/marx/works/1852/18th-brumaire/ch01.htm

Meadowcroft, J., 2009. What about the politics? Sustainable development, transition management, and long term energy transitions. *Policy Sciences*, 42 (4), 323–340. doi:10.1007/s11077-009-9097-z

Moore, J.W., 2000. Sugar and the expansion of the early modern world-economy: commodity frontiers, ecological transformation, and industrialization. *Review (Fernand Braudel Center)*, 23 (3), 409–433.

Mouffe, C., 1979. *Gramsci and Marxist theory (RLE: Gramsci)*. London: Routledge.

Newell, P. and Mulvaney, D., 2013. The political economy of the 'just transition'. *The Geographical Journal*, 179 (2), 132–140. doi:10.1111/geoj.12008

Norval, A., 1997. Frontiers in question. *Filozofski vestnik*, 18 (2), 51–75.

Perkins, H.R., 2011. Gramsci in green: neoliberal hegemony through urban forestry and the potential for a political ecology of praxis. *Geoforum*, 42, 558–566. doi:10.1016/j.geoforum.2011.05.001

Perreault, T., 2013. Dispossession by accumulation? Mining, water and the nature of enclosure on the Bolivian Altiplano. *Antipode*, 45 (5), 1050–1069. doi:10.1111/anti.12005

Rockström, J., *et al.*, 2009. A safe operating space for humanity. *Nature*, 461, 472–475. doi:10.1038/461472a

Roger, S., 1991. *Gramsci's political thought: an introduction*. London: Lawrence & Wishart.

Ruz, C., 2011, October 31. The six natural resources most drained by our 7 billion people. *The Guardian*. Available from: www.theguardian.com/environment/blog/2011/oct/31/six-natural-resources-population

Saull, R., 2012. Rethinking hegemony: uneven development, historical blocs, and the world economic crisis. *International Studies Quarterly*, 56 (2), 323–338. doi:10.1111/j.1468-2478.2012.00720.x

Schilling-Vacaflor, A., Flemmer, R. and Hujber, A., 2018. Contesting the hydrocarbon frontiers: state depoliticizing practices and local responses in Peru. *World Development*, 108, 74–85. doi:10.1016/j.worlddev.2018.03.019

Scott, J.C., 1998. *Seeing like a state: how certain schemes to improve the human condition have failed*. New Haven, CT: Yale University Press.

Shaw, L.E., 2013. Asteroids, the new Western frontier: applying principles of the general mining law of 1872 to incentive asteroid mining. *Journal of Air Law and Commerce*, 78 (1), 121–169. Available from: https://scholar.smu.edu/jalc/vol78/iss1/2

Steffen, W., *et al.*, 2015. Planetary boundaries: guiding human development on a changing planet. *Science*, 347 (6223), 736. doi:10.1126/science.1259855

Thomas, P.D., 2009. *The Gramscian moment: philosophy, hegemony and Marxism*. Leiden, Netherlands: Brill.

Turner, F.J., 1958. *The frontier in American history*. New York: Henry Holt and Company.

Van Teijlingen, K. and Hogenboom, B., 2014. *Development discourses at the mining frontier: Buen Vivir and the contested mine of El Mirador in Ecuador* (ENGOV Working paper No. 15). Available from: https://pure.uva.nl/ws/files/2438349/160268_447910.pdf

Veltmeyer, H., 2000. Post-Marxist project: an assessment and critique of Ernesto Laclau. *Sociological Inquiry*, 70 (4), 499–519. doi:10.1111/j.1475-682x.2000.tb00922.x

Vorbrugg, A., 2019. Ressourcen. *In*: J. Brunner, A. Dobelmann, S. Kirst and L. Prause, eds., *Wörterbuch Land- und Rohstoffkonflikte*. Bielefeld: Transcript, 272–278.

Watts, M.J., 2018. Frontiers: authority, precarity, and insurgency at the edge of the state. *World Development*, 101, 477–488. doi:10.1016/j.worlddev.2017.03.024

Weber-Fahr, M., 2002. *Treasure or trouble? Mining in developing countries*. Washington D.C.: World Bank and International Finance Corporation.

Winkler, H. and Marquand, A., 2009. Changing development paths: from an energy-intensive to low-carbon economy in South Africa. *Climate and Development*, 1 (1), 47–65. doi:10.3763/cdev.2009.0003

Woodworth, M.D., 2017. Disposable ordos: the making of an energy resource frontier in western China. *Geoforum*, 78, 133–140. doi:10.1016/j.geoforum.2016.04.007

Yakovleva, N. and Vazquez-Brust, D., 2012. Stakeholder perspectives on CSR of mining MNCs in Argentina. *Journal of Business Ethics*, 106 (2), 191–211. doi:10.1007/s10551-011-0989-4

Žižek, S., 2000. Class struggle or postmodernism? Yes, please! *In*: J. Butler, E. Laclau and S. Žižek, eds., *Contingency, hegemony, universality: contemporary dialogues on the left*. London: Verso, 90–135.

3 Fossil Fuel Dependency in South Africa

Fossil-fuel hegemony is a historical process that favours fossil-fuel exploitation and use over other energy sources by foreclosing alternatives such as renewable energy. The growing use and exploitation of fossil fuels result from a spatial expansion of the fossil-fuel frontier, causing a struggle between hegemonic actors and resisting civil society actors mostly in places of extraction. Fossil-fuel hegemony is sustained by discourses revolving around progress, cheap energy supply, job creation, and abundance against the backdrop of historical commitments to fossil-fuel energy and its cultural embeddedness. Especially in global so-called emerging economies, the number of planned fossil-fuel projects surges. However, the 'fossil bloc' is challenged by civil society actors (Goodman *et al.* 2020).

In this chapter, I first give an overview of fossil-fuel extraction in South Africa from industrialisation to the end of apartheid. I show that fossil-fuel hegemony relied on the interlinkages between government, industry, and the white population's persuasion by cheap energy provision under apartheid. I characterise mining relations as racialised projects still impacting the black majority population the worst. Environmental NGOs during apartheid were very narrowly focused on conservation spaces and did not challenge the environmental damage of the fossil-fuel industry in poor, black neighbourhoods. In the second section, I describe post-apartheid fossil-fuel politics, emphasising the continuing impacts of big lobby groups on ANC government policy-making and continuities of the 'historical bloc' sustaining fossil fuels. There has even been a backlash towards more centralisation, expansion, and fast-tracking fossil-fuel projects more recently. In the last section, I look closely at the legal provisions pertaining to the inclusion of civil society actors. ANC mining policy-making withdrew certain checks and balances introduced after apartheid in the name of fast-tracking development.

Emergence and Consolidation of the Historical Bloc Around Fossil Fuels

From the start of industrialisation in South Africa in the late 19th century, coal has played a central role in the economy. The industrial-scale gold rush towards the end of the 19th century was hungry for energy supply from coal. Industrialisation

DOI: 10.4324/9781003110835-4

led to the rapid spatial expansion of coal mines and the exploitation of the vast coal belt stretching from south of Johannesburg to the northern parts of KwaZulu-Natal. Coal mining also gave birth to mining towns such as Witbank to the west of Johannesburg (renamed as Emalahleni in 2006, which translates to 'place of coal'). Other small towns such as Newcastle grew rapidly due to industrial-scale coal mining.

From the late 19th century, the expansion of coal mining has severe consequences for life in the vicinity of mines and beyond. Coal mining set in motion the proletarisation and ghettoisation of black workers and their families in the most polluted and unfertile lands (Saul and Bond 2014; Beinart 2001). From its inception, industrial mining in South Africa was a racialised class project furthering the interests of white mining capital at the expense of the black majority population. A cheap black labour system was sustained by introducing a migrant workforce from neighbouring countries in order to 'divide and rule' the black working class. The ideologies of racial superiority of the white minority justified this rule by force. Black workers were denied managerial jobs in the mines by the 'colour bar', which legislated racial superiority from the 1920s (Davenport 2013, p. 354).

Mining businesses created a new elite working class at the intersection between private mining business and politics. Cecil Rhodes, for example, was both the founder of de Beer's mining venture and Prime Minister of the Cape Colony at the end of the 19th century. This trend marked the beginning of the entrenchment of mining interests in public affairs. As early as 1887, mining houses established the powerful lobby group, South African Chamber of Mines, to ensure 'the promotion and protection of the mining industry' (Munik 2010, p. 4). Fine and Rustomjee (1996) termed the close interlinkages between mining capital, the energy sector, and the political elite that characterised the country's political economy from industrialisation, the minerals-energy complex (MEC). In addition, colonial and apartheid regimes assigned a key role to traditional black chiefs by co-opting them into the MEC. The chiefs' roles in rural black areas were strengthened to gain their nominal approval for mining and suppress community protests (Manson and Mbenga 2014).

Coal mining played a central role in supporting the gold mining boom, starting with the discovery of gold and diamonds. Gold mining companies owned many coal mines. They used cheap coal as an input to increase revenue from gold rather than to profit from coal directly (Munik 2010, p. 4). The 1920s saw an expansion of the coal-fired energy market in places like Witbank in today's Mpumalanga province. The state-owned utility Escom (today's Eskom), established in 1923, began to electrify middle-class white households in coal mining areas, winning them over to endorse a coal-fired transformation. In the early 20th century, pollution in coal mining areas such as Witbank sparked early environmental criticism. Accounts of farmers show that they opposed coal mining in anticipation of pollution and disturbances to land. However, local administrations gave preferential treatment to early coal mining companies and facilitated coal mining expansionism (Singer 2011, p. 286).

In the 1920s, the city of Witbank even advertised the pollution from coal, praising the 'smoky horizon, dusty red sunsets, smoke stacks and black mountains of coal waste' (Singer 2011, p. 289). While the narrative of hazy skies did not particularly persuade most residents, the state-owned utility Eskom defended the necessity of large-scale coal mining by diverting attention towards the advantages of electricity. In fact, for white middle-class households, the choice between open fires in the houses versus the convenience of electric appliances, including electric heaters, supported the arguments in favour of 'progress'. The utility company managed to make an argument 'entrenching the association between electricity and modernity' (Singer 2011, p. 290), of which coal mining was a necessary – albeit unloved – prerequisite. In Singer's words, 'the environment became a necessary sacrifice for the new beauty of modern living' (2011, p. 291).

At critical political junctures, the industry lobbied successfully for the expansion of fossil fuels. Coal mining was repeatedly in crisis in South Africa. Shortages in the power supply at the end of the 1960s were met with increases in the power supply from coal and the construction of more coal-fired power plants (Vermeulen 2015). The 1973–1974 oil crisis opened a window of opportunity for exports and made South Africa a cheap coal exporter (Eberhard 2011, p. 7). South African coal is abundant compared to most other countries but of lesser quality. Taverner-Smith, taking up his professorship of Mineralogy in 1973 described in his inauguration lecture the properties and potentials of South Africa's coal.

> South African coal is inferior in quality to European or American coal, but at least we have it in abundance. It has been estimated that coal mined in this country up to the present time totals about 700 million tons, but conservative estimates place known reserves at about 12 000 million tons. This would be enough to sustain the country for more than 200 years, even at the present high level of consumption. Long before that, it is anticipated that atomic or solar energy will have superseded (sic.) both coal and oil as the principal source of power.
>
> (Taverner-Smith 1973, pp. 9–10)

Taverner-Smith remarks how the expansion of coal mining remained relatively unnoticed by the general public. He regarded coal as sharing a similar fate to Cinderella: 'commonly overlooked in favour of more glamorous minerals' (1973, p. 15). He said that:

> South Africa is probably more dependent on coal than any other industrialized nation. Yet in comparison with gold, diamonds and other 'glamour' minerals like platinum or chromite, we hear very little about it. All our electricity, including all the power for heavy industry, is generated from coal, and we turn lower grade coal into oil, liquid fuels and chemicals to a greater extent and more cheaply than anywhere else in the world. Coal is undoubtedly of fundamental importance to us.
>
> (Taverner-Smith 1973, p. 1)

South African coal became a global commodity when the first coal-export terminal was built in Richards Bay in 1976 to capitalise on export opportunities with Japan and later several other countries around the world (Munik 2010, p. 4). Richards Bay is well placed to conveniently serve both Eastern and Western hemispheres (Eberhard 2011). With growing energy demand in the 1970s, the government explored the gas reserves in the semi-arid desert of the Karoo. However, the resource was deemed unrecoverable as the gas was 'too tight' in the rock (Scholes *et al.* 2016). With fracking techniques, the issue would resurface four decades later.

The 1970s saw an expansion of coal mining in South Africa, but the expansion of coal mines also caused the deterioration of the environment. The Chamber of Mines admitted disturbances to the environment caused by mining, claiming that it would only temporarily obstruct agriculture. Three arguments were put forward in favour of coal exploitation: first, its compatibility with other land uses and the environment (as a reaction to early environmental criticism); second, its cost-effectiveness; and finally, its abundance. A report on *Rehabilitation of Land Disturbed by Surface Coal Mining* published by the Chamber of Mines in 1981 discusses the environmental impacts of mining.[1] This report admits that surface coal mining (a relatively novel technique) causes significant damage to the environment. Still, the authors claim that these disturbances could be mitigated, and hence, environmentalism and mining can be reconciled. Environmental criticism of coal mining was incorporated into the narrative of the Chamber of Mines, presenting coal mining as being only temporarily obstructive and, therefore, compatible with land uses such as agriculture.

> Surface mining necessitates alienation and radical disturbances of land. Given that there is a growing sense of concern at encroachment on agricultural land by urban/industrial spread and burgeoning transportation networks, surface mining has the potential to become a sensitive environmental issue, notwithstanding the essential nature of its end product. Fortunately, however, if proper *land rehabilitation* follows in the wake of mining, then the alienation of land need be no more than temporary (unlike most other kinds of alienation), reduction in the utility of land need no more than slight and the environmental effects imperceptible. Mining and agriculture are not necessarily mutually exclusive except over a relatively short period of time.
>
> (Chamber of Mines 1981, p. iv; emphasis in original)

However, the report highlights that the rehabilitation of mines should also be profitable in the sense that rehabilitation should not follow 'unreasonably high standards' (Chamber of Mines 1981, p. v). This puts the earlier discussion in a different light: rehabilitation should only mitigate land disturbances as long as they are cost-effective. South Africa's competitive advantage is the low-cost coal production, which conflicts with 'unreasonably high standards' in environmental mitigation.

In 1983, W.W. Malan, the then president of the South African Chamber of Mines, announced that: 'the future of South Africa's coal industry remains sound. The country is a low-cost producer with large reserves of coal' (Chamber of Mines

1983, p. 36). Indeed, the 1980s saw peaks both in employment (in 1987) from mining as well as mining's contribution to national GDP (in 1980). Economic recession towards the end of the 1980s brought the apartheid government to the brink of collapse. In the same period, revenues from mining and employment plummeted. While the debt crisis and falling exports caused a serious economic crisis, critique of environmental devastation from mining remained peripheral. White environmentalists focused on nature conservation during apartheid, which was in line with the regime's idea of having a green veneer. Most white environmental groups did not challenge apartheid's accelerating pollution in black neighbourhoods but agreed on safeguarding the environment in fenced conservation spaces funded by the government. This amounted to a passive endorsement of the fossil-fuel hegemony among the environmentally conscious white population:

> the non-governmental sector of the South African environmental movement continued to focus predominantly on the conservation of fauna and flora, and of particular areas that were fenced in to ensure the continuation of their existence. These protected areas become symbols of responsible stewardship of the natural environment for the South African government, the National Parks Board, the provincial nature conservancies, a number of ENGOs and a large segment of the white people in the country.
>
> (Steyn 2005, p. 394)

As long as the fossil-fuel frontier did not interfere directly with the interests of conservation areas, green NGOs did not campaign against pollution. While the white minority population was thus incorporated in state policy, the organised black opposition to apartheid was not primarily concerned with environmental policy. Rather, they desired service provisions such as electricity connection, which was available to only 50% of South African households until 1994 (Wilkinson 2015). The organised and environmentally conscious part of white civil society was primarily concerned with conservation. Black civil society groups were fighting a political battle to topple the racist apartheid state. Hence, there was relatively little focus on environmental critiques of fossil-fuel hegemony. Towards the end of apartheid, the Environmental Justice Network Forum (EJNF) was the first nationwide environmental group to consider the environmental ramification of South Africa's industries. EJNF was also the first group to formulate a coherent climate change critique against environmental pollution from big industries (Leonard 2011). They started asking 'why poor people live in poor environments' (Leonard 2018, p. 25), but most environmentally conscious civil society groups did not incorporate human aspects of environmental destruction into their politics. Green NGOs under apartheid focused almost exclusively on the non-human environment and collaborated with the apartheid state.

To sum up the main tenets of fossil-fuel hegemony until the end of apartheid, I described a 'historical bloc' between state, business, and civil society hegemonising fossil fuel use and exploitation by building a 'common sense' (cultural, material, and historical) around their project. In analyses of the long survival of apartheid, South Africa's resource abundance is often forgotten, and the role fossil fuels played to entrench political power. Under apartheid, the deepening of exploitation of fossil fuels was legislated and implemented by entrenching white privilege and suppressing the black majority population. White households enjoyed cheap electricity, and the black majority were forced to live in the most polluted parts of the country. As seen early in South Africa's mining history, coal was rendered equivalent to modernity and rolled out cheaply to white middle-class households. State-owned utility Eskom is an example of the 'integral state' that entrenched fossil fuel use through the state's active promotion of coal mining. The hegemonic position of fossil fuels became particularly visible when the political opportunity of the oil crisis in 1973–1974 was exploited to entrench coal mining in South Africa. This leadership role of the MEC was hardly challenged, as environmental activism was geared towards conservation spaces. Starting in the 1970s and the 1980s, black social movements resisting the apartheid system were '[t]he trade union movement, civic associations, youth groups, women's organizations and professional groups' (Shubane 1992, p. 35). However, these were primarily concerned with ending a racist polity rather than tackling environmental issues directly (Leonard 2011, p. 115).

Post-Apartheid Fossil-Fuel Politics

After 1994, with few concessions, the 'historical bloc' around fossil fuels in South Africa remained intact under the ANC. Originally, in economic policy, the ANC had certain radical economic policy changes in mind. Shortly before Mandela was released from prison in February 1990, he announced that 'the nationalisation of mines, banks and monopoly industries is the policy of the ANC and a change or modification of our views in this regard is inconceivable' (Bond 2014, p. 13). That same year, many ANC cadres such as Jacob Zuma, Thabo Mbeki, and Joe Slovo returned to South Africa from exile, preparing for the first democratic election campaign. The ANC spelled out its plans in the policy guidelines *Ready to Govern*. The document pledged that the ANC would work towards an equitable society, crucially highlighting that poverty and environmental degradation were closely related problems caused by the apartheid system (Steyn 2005, p. 397). Key rallying points of the ANC programme were to deliver basic services such as water, housing, and electricity. However, when the ANC came to power, they were faced with a macro-political environment in which neo-liberalisation was an almost universal dogma. The strained budgetary situation they inherited from their predecessors as well as global norms of fiscal prudence soon caused them to depart from more redistributive policy-making. With the ANC's former allies in the Soviet Union conceding defeat, the ANC's policy-making underwent major influence by World Trade Organization (WTO) and World Bank standards and thinking. For example, first Trade and Industry Minister Trevor Manuel eagerly facilitated lowering import barriers for trade with Europe and the US overfulfilling WTO standards. Especially the textile and automobile industries suffered disproportionately (Bond 2014, pp. 30, 39). According to Bond (2014), South Africa saw an 'elite transition' that left property relations intact and created a very small black bourgeoisie becoming part of, not altering, the MEC.

Much like prior decision-makers during apartheid and before, post-apartheid leaders are very close to mining interests (Finkeldey 2020). Cyril Ramaphosa left public service in the late 1990s to become one of the richest South Africans with equity in some South African companies, including mining businesses (Marinovich 2016, p. 134). On his return to ANC frontbench politics, Ramaphosa raised suspicion due to his role as a shareholder of the mining company Lonmin in 2012. Emails sent to the mining management and ANC elites are regarded as contributing to a climate in which the Marikana massacre took place on 16 August 2012, killing 34 striking miners (Marinovich 2016, p. 137). Only four months after the massacre, Ramaphosa was promoted to Vice-President of the country.

Ramaphosa's predecessor Jacob Zuma left the presidential office prematurely because of corruption charges and allegations of undue influence given to third-party business interests. During Zuma's terms in office, his close kin notoriously received government positions, and the family businesses raised serious concerns over conflicts of interests between public office and private investments (City Press 2014). Zuma's nephew Clive Khulubuse Zuma is implicated in the offshore scandals revealed in the Paradise Papers, offshoring oil revenues from the Democratic Republic of Congo in tax havens (International Consortium of Investigative Journalists 2016). Jacob Zuma granted the Gupta family, which is heavily invested in South African mining, what was established to be undue influence over government decision-making (Madonsela 2016). His successor Cyril Ramaphosa assumed office on a zero-tolerance for corruption ticket after a raid on the Gupta family's home sealed the latter's departure (BBC 2018). Ramaphosa changed energy policy to a degree by scrapping Zuma's plans for new nuclear plants. The fossil fuel agenda of the ANC remains in place, however. Ramaphosa renewed his commitment to mining as the 'sunrise industry', highlighting the 'massive potential for South Africa's mining industry to grow, create jobs, stimulate industrial activity and promote social development' (Mining Review 2018). The mining industry enjoys the full support of the current administration. Ramaphosa reassured traditional authorities and the Zulu King Zwelithini that they would continue to have a stake in decision-making on land questions (Mahlase 2018).

Especially in KZN, the role of traditional authorities in land questions remains particularly powerful. One-third of the land is under the formal custody of King Zwelithini, who praises mining as 'a major creator of jobs in rural areas' and has granted numerous mining authorisations since his trust, the Ingonyama Trust, began administering land in 1993. The King is a crucial gatekeeper in expanding the fossil-fuel frontier under tribal control (Harper 2015). According to Zwelithini, traditional chiefs (ama Khosi) have an important role in mining, as they 'should be the drivers of mining in rural development' (Harper 2015). Rather than 'tilling of the soil', Zwelithini's vision is to transform local livelihoods into spaces of the modern mining industry.

The role of pro-fossil-fuel lobby groups has also expanded under ANC rule. The pressure of big lobby groups such as the Chamber of Mines, which was established at the birth of industrial mining, was complemented by new groups such as the Energy Intensive User Group (EIUG) of Southern Africa in 1999. EIUG is successfully advocating low energy prices and low wages for workers. The same year, Coaltech Research Association was formed, advocating 'the coal industry for the national interest' (Coaltech n.d.). Numerous links can be traced between the state-owned utility Eskom, Eskom's holdings, and private enterprises (Greenberg 2009). For example, in 2001, on the First Annual Banquet of the Fossil Fuel Foundation (FFF) of Africa, the Prestige Award went to Ian McRae, former CEO of Eskom (FFF 2018). Organisations such as FFF and Coaltech arrange events to collaborate between industry, government, and academia, fostering a 'common sense' around fossil-fuel exploitation and use.

In line with these lobby groups, South African policymakers highlight fossil fuels as the only viable source to meet the economy's demands and the population. The ANC in government made it a rallying point in their political campaign to electrify poor households and keep energy prices low. A total of 85% of South African households were connected to the grid by 2012 (Statistics South Africa 2013, p. 25). This successful service provision was sustained by the world's lowest energy prices for consumers (Moolman 2017). According to the South African Department of Energy (DoE), 'coal is abundant, affordable, easy to transport, store and use, plus free of geopolitical tensions; all these attributes make it very popular' (DoE n.d.). The department also highlights a 'relative lack of suitable alternatives to coal as an energy source' (DoE n.d.).

However, an energy crisis put the fossil-fuel hegemony to the test in 2008 and some of the following years. The crisis was met by commissioning two of the world's largest coal-fired power plants. Co-funded by foreign investors, two of the world's largest coal-fired power stations were put under construction. Madupi power station in Witbank is backed by the biggest ever World Bank loan ($3.05 billion) (World Bank 2010), and Kusile power station in Limpopo close to Botswana's border is backed by a $2.5 billion Chinese government loan (Khumalo 2018). These power stations are operated by state-owned utility Eskom, which remains almost exclusively reliant on power from coal. With growing pressures on coal as the dirtiest source for energy generation, technological fixes are emphasised to transform coal into a 'clean' energy source. For example, the new coal-fired Kusile power plant is advertised as 'one of the cleanest coal-fired power plants' in Africa (General Electric 2017). This shows that the main tenants of the hegemonic discourse on abundant, clean, and affordable coal have not changed over 30 to 40 years. As in the 1970s, an energy crisis was exploited to entrench fossil fuel use even further (Vermeulen 2015).

The massive expansion of coal is complemented by plans to use hydraulic fracturing methods to extract gas from rock formations deep below the surface. Fracking in South Africa was first proposed in 2011. The ANC sustains that hydraulic fracturing could create more than 700.000 jobs (upstream and downstream) and presents it as an instrument for poverty alleviation and 'game-changer' for the economy. This rhetoric is used to highlight the party's readiness to aid historically disadvantaged South Africans. Deputy Minister of Mineral Resources Godfrey Oliphant has frequently used the promise of 700.000 jobs to promote the idea of hydraulic fracturing in South Africa. The number originates from a Shell-funded study into the economic potentials of fracking (Econometrix 2012). In the parliamentary debate on fracking on 1 December 2016, ANC MP Imamile Pikinini was even more unambiguous about the role of corporate-funded studies for the ruling party:

> A study commissioned by Shell estimates 50 trillion cubic feet of potential reserves which would add $20 billion to the GDP of South African economy every year for 25 years and create 700.000 jobs. The same report abounds with evidence of number of multinational corporations that have interests to invest in the shale gas sector in South Africa to the great benefits of our domestic socioeconomic development.
>
> (FFSA 2016)

This statement suggests a considerable influence of oil and gas companies on the ANC executive. Environmental activists consider this number highly inflated and point to job losses in agriculture where fracking would take place. Internal debates in the biggest opposition party, Democratic Alliance (DA), controversially discussed fracking, as some members were worried about the environmental repercussions of the drilling technique. However, fracking was finally supported, mainly because of its potential to create new jobs (interview with politician, 15 February 2017).

The entrenchment of fossil-fuel hegemony across party benches is coupled with the acceleration of energy projects. South Africa's government has repeatedly insisted on its readiness to 'cut red tape' in the extractive sector and fast-track mining developments. Emblematic of the government's willingness is 'Operation Phakisa', launched in 2014 (which translates to 'Hurry up' in Sesotho). The operation aims to bring about a 'new spirit of moving faster in meeting government's targets' (Department of Environment n.d.). Phakisa is often invoked by ANC politicians, including President Ramaphosa, underlining the willingness of the government to develop the country's resources more efficiently and create a sense of urgency, normalising fossil-fuel extraction. The methodology, however, remains opaque. The government has not published documents or policy papers discussing concrete mechanisms or processes to implement fast-tracking at the time of writing.

Under the ANC, expanding the fossil-fuel frontier and other mining practices is presented as being in the national interest, giving the country a competitive advantage over other African countries. South Africa is one of the major global coal exporters, and behind China, the USA, India, Australia, and Indonesia, it is the sixth biggest coal producer. Multinational companies such as AngloCoal, BHP Billion, and Glencore are responsible for the majority of coal production. Nineteen coal fields are currently recognised in South Africa, covering almost ten million hectares (Hancox and Götz 2014, p. 19). Therefore, the DMR boasts that 'South Africa is a heavyweight in global mining'.

However, increasingly, the MEC runs into contradictions as environmental degradation and skyrocketing consumer prices heightened discontent. Between 2007 and 2017, the energy price rose by 350%, significantly above the rate of inflation (Moolman 2017). Also, the ANC had a hard time dealing with the employment crisis in the country. With higher electricity prices, the accelerated environmental and social costs of fossil-fuel expansion called into question the 'common sense' of fossil fuels. The big international NGOs are pointing to the unsustainability of South Africa's extractive model. Oxfam, Greenpeace, and ActionAid call on the South African government to diversify the energy mix instead of relying almost exclusively on cheap coal. These NGOs are pointing to the abundant potential for solar power and other renewables. For example, Greenpeace considers South Africa the third-best-suited country for solar power installation (Mail & Guardian 2015). Greenpeace presents a 'practical blueprint for cutting carbon emissions – replacing coal and nuclear fuels with renewable energy (such as sun and wind) and energy efficiency' (Greenpeace 2011, p. 5). The disconnection between the bigger-picture critiques of international NGOs and local NGOs' day-to-day challenges to fossil-fuel expansion remains a dent in the environmental movement in South Africa. Notoriously, the environmental movement in South Africa, as elsewhere in the world, suffers from a failure to connect the dots (Bond and Friends 2010; Cock 2006). The ANC has conceded occasionally to pressure from social movements. For example, when after civil society protests, the Vaal Triangle was declared an air pollution 'hot spot' (Steyn 2005, p. 399).

Regulatory Framework

The ANC in government 'struggled to come to terms with the environmental legacy of the apartheid era' (Steyn 2005, p. 398), having to find ways to undo century-long injustices. On the whole, the ANC made some incremental steps to protect the environment better. After the ANC came to power, the first democratic constitution in 1996 enshrined the right to a healthy environment. However, the latest developments suggest that the ANC is retreating from its commitments to environmental protection.

This section will outline how legal provisions in post-apartheid South Africa are characterised by centralisation and selective participatory governance mechanisms. I will also show that there are severe shortfalls concerning environmental oversight and enforcement of environmental protection. The Department of Mineral Resources (DMR) is notoriously understaffed, having at the same time assumed more discretionary power over environmental protection in the last few years. Timeframes during which appeals can be made were significantly shortened. Legal provisions are geared towards the interests of the MEC while notionally guaranteeing the participation of the citizenry in decision-making. Hegemony is exercised by the selective inclusion of civil society actors in prescribed stakeholder arrangements.

By law, every endowment below the surface belongs to the South African government. Only the government can issue mineral rights (Hedden *et al.* 2013, p. 6). The major legislation that applies to the mining sector is the Minerals and Petroleum Development Act of 2002 (MPRDA). The legislation provides that mining companies must apply to the Minister of Mineral Resources for a licence to operate. Rights granted are registered with the Mineral and Petroleum Titles Registration Office. Based on gross sales, rights holders are obliged to pay royalties to the National Revenue Fund. Before a right is granted, the mining right applicant must agree on a Social and Labour Plan (SLP) that will be subjected to an annual review. If the SLP is not adhered to, the Department of Mineral Resources can revoke the mining right. SLPs can be amended if the Minister agrees to the changes. However, the reviews of the minister are not publicly available and the powers of the department 'are not always exercised' (Mandela Institute 2017, p. 20). The law is discretionary about the agreement between the applicant and the department. The MPRDA does not require the government to publish mining rights agreements, but companies must consult the affected communities before a mining right comes into effect (Mandela Institute 2017, pp. 20–21).

Environmental provisions feature prominently in South African law. Crucially, section 24 provides that every citizen has the right to an environment that is not harmful. The National Environmental Management Act, 107 of 1998 (NEMA), determines that applicants must fulfil the requirements of an environmental impact assessment (EIA) that assesses the environmental risks of the applied practices and agrees on measures to mitigate impacts. Registered interested and affected parties (I&APs) are given 30 days to comment on proposed developments (Fig 2012, p. 22). Provisions under the NEMA oblige applicants pursuing mineral exploitation to undertake EIAs. Comments made in the process of the application must be recorded and forwarded to the relevant authorities. The EIA is there to inform I&APs about the impacts of the respective land-use proposals. Applications have to undergo different stages that have to be applied for separately (scoping, exploration, and exploitation). Successful applications for large-scale extraction projects take years to complete. On the one hand, this leaves ample opportunities to stop mining developments at different stages of the process. On the other, the processes can be so long lasting that citizens' action challenging extraction is hard to sustain.

NEMA allows mine-affected communities to respond to the Department of Mineral Resources. It has been debated against the backdrop of global norms of ecological modernisation and liberal democratic ideals of participation. The framework has been criticised for its selective inclusion of certain civil society stakeholders, making the EIA process almost exclusively an expert forum (Barnett and Scott 2007). Even participants whose first language is English have lamented the technical nature of the EIA process (interview with activist, 22 October 2016).

The EIA practitioners I interviewed for this study emphasised that they believed that the procedure itself is sound. But the implementation of mitigating measures – a condition for the granting of environmental authorisation – is often lacking. An environmental practitioner and frack-free activist commented on the EIA process:

> there is room for appeals and interdicts and that kind of thing. [The EIA, sic.] should be subject to certain conditions, and I've seen many examples of how those are just not adhered to, and there is very little enforcement and compliance and monitoring.
>
> (interview with activist, 26 January 2017)

The activist lamented that 'there seems to be very little ability and capacity with the departments to really keep a tight rein on the compliance issues' (ibid.).

Some of the constituents who would be affected by mining are not informed about the proposed land use. In areas where there are big mining applications for hydraulic fracturing, the circulation of information is insufficient. The EIA procedure is geared towards English-speaking, internet-savvy, formally educated forms of engagement. Despite there being translators into indigenous languages, at times, access to information is insufficient. Some of the constituents who would be affected by mining are not informed about the proposed land use.

From a Gramscian perspective, these consultations provide ways to incorporate civil society actors into the hegemonic bloc or at least de-radicalize their claims. Mining companies facilitate consultations on extractive projects. These consultations do not take place on a 'level playing field', as companies choose and pay for the consultants that manage their projects. By design, the relationship between consulting companies and consulted civil society is such that consultation often translates to giving consent. The dissent voiced at public consultations is noted in final reports and taken into consideration by state regulators from departments actively promoting mining. There is also a lack of resources at the government level to manage all mining projects and ensure that they follow due procedure. Leonard (2016, p. 16), commenting on public participation in the mining sector in South Africa, observes:

> Despite the advent of democracy witnessing government making considerable progress in developing the legal framework to manage existing and new mining operations and to include citizens in decision-making processes, government has engaged in tokenistic participation and has not effectively monitored mining development or included citizen concerns into decision-making processes. Empirical results suggested a number of complexities and factors influencing poor governance and participation. Lack of human resources at all government levels contributed to ineffective monitoring of mining developments, suggesting a continuation of apartheid ineffectiveness of enforcement processes although with varying degrees.

The lack of resources and the partisanship of government departments contribute to a distrust of activists towards consultation processes. There is considerable debate as to whether the institutional participatory framework is robust enough to ensure accountability of applicants and proper consultation of I&APs

(Mandela Institute 2017; Leonard 2016; Centre for Environmental Rights [CER] 2015; Friedman 2006). As protest events around fossil-fuel projects show, formal participatory channels cannot entirely absorb dissent against fossil-fuel projects.

Decisions to grant or revoke licences can only be taken at the national level but mostly follow the advice of regional managers in the provinces. A staff member of the regional DMR office in Durban criticised his department's lack of resources to oversee mining companies. Employees of the department cannot attend to all conflicts between mining companies and mine-hosting communities (interview with staff member of Department of Mineral Resources, 13 April 2016). The capacity of the regional DMR to settle conflicts between companies and communities or comprehensively monitor environmental harm is lacking. There are, however, institutionalised grievance mechanisms that can be addressed to government. The Regional Mining Development and Environment Committee (RMDEC) is a non-binding settlement procedure attended by different government departments that collect stakeholders' arguments against or in favour of certain developments. Every group wishing to speak on the occasion is invited to voice their grievances in the presence of the regulating authorities (CER 2014, pp. 10–12).

The national environmental legislation was amended in 2014, providing an 'unusually complex law' that places the environmental oversight over mining operation with DMR (CER 2015). Since then, NEMA authorisations 'relating to prospecting, mining, exploration or production' have fallen under the discretion of DMR (Humby 2015, p. 125). This law has been commonly referred to as the 'One Environmental System' and passed to streamline the monitoring and compliance system. The 'One Environmental System' shortens the timeframes for both environmental oversight as well as public participation, casting doubt over the two constitutional requirements for sound environmental assessment and participatory democracy (Humby 2015, p. 130). Precedent cases have shown that DMR did not stop the 'illegal activities' of mining companies (CER 2015). Conservationists, national and international environmental NGOs, and the CER in Cape Town raised serious concerns prior to the bill being signed into law (CER 2013). In KZN province alone, the Department of Economic Development, Tourism, and Environmental Affairs (EDTEA) holds 11 district offices (2014). DRM in KZN only operates from one regional office in Durban. This issue stands in inverse relation to the actual powers of the respective departments.

Prior to the 'One Environmental System', environmental monitoring and compliance were the responsibility of the Department of Environmental Affairs (DEA). Nevertheless, the DEA can still declare entire areas unsuitable for mining and hence pre-empt any application in these declared pockets (Humby 2015, p. 130). At the provincial level, the DEA remains a potential veto player to mining projects if high-priority conservation areas are potentially impacted. In KZN, there is an increasingly less funded biodiversity section within the provincial DEA known as Ezemvelo Wildlife, which has been quite vocal about certain mining projects (Breed 2016). Ezemvelo Wildlife are mandated to protect high-biodiversity areas in the province and automatically comment on development applications potentially affecting these areas. This shows both the potential muscle of Ezemvelo's engagement as well as the limitations of their mandate. If an area has already been earmarked as a preferential biodiversity area, Ezemvelo's opposition to DMR can hardly be ignored.

Municipalities have to sign off mining applications at the local level, which gives them some de facto power to decide based upon the merits of applications. In the last instance, however, as a municipal development planner at the Amajuba District Municipality explained: 'If the minister [of Mineral Resources, sic.] overrides everything and gives them the go-ahead to mine . . . there is nothing we can do about it, unfortunately' (interview with staff member of district municipality, 1 November 2016). However, in cases in which municipalities mapped their preferred land uses, companies sometimes refrained from launching applications for mineral extraction in anticipation that their applications would be turned down (ibid.).

South Africa's multi-level governance in the mining sector is marked by centralisation and streamlining of procedures. Piper and von Liers (2016, p. 316) even detect a 'failure of formal participatory governance in South Africa due to poor institutional design and de facto capture into informal political logics'. As shown, public participation is particularly geared towards incorporating an active part of civil society actors. Consultations are conducted and geared towards informed middle-classes, effectively excluding non-English speaking people who will have difficulties following the process. To conclude, public participation processes in the extractive sector amount to containing and depoliticising highly contentious issues. Activists often lament that consultations alone are taken as evidence of their active consent to mining ventures.

Conclusion

In this chapter, I have discussed the 'historical bloc' since industrialisation through apartheid and its consolidation in post-apartheid under the ANC. The 'historical bloc' around fossil fuels between government, corporations, and the white minority population as consumers under apartheid was sustained by traditional black chiefs coercing 'their' communities into mining. Colonialism and apartheid entrenched a regime that relied heavily on the exploitation and spatial confinement of the black majority. Colonialism and apartheid's social and spatial engineering practices still have a lasting and lingering impact on today's human geography, income distribution, and social status. With mining also came a new class of mining magnates holding public office and building a fortune in the mining sector.

The ANC came to power on the expectation that the regime would change direction significantly. Reflecting on the mining industry, President Ramaphosa said: 'Apartheid was baked hard in the mining industry because that's where it originated' (Marinovich 2016, p. 131). The ANC directed their focus on mining even more than the previous regime did. The party promised to pull the black masses out of poverty and electrify poor households without significantly departing from the almost exclusive reliance on coal. In contrast to the apartheid system, the ANC persuades black voters by rolling out the grid to historically discriminated masses. To expand the cheap energy supply, state-owned utility Eskom remained heavily reliant on coal lobbied by fossil-fuel-funded organisations such as the Energy Intensive User Group (EIUG) of Southern Africa. The aims of these lobby groups are in tension with the country's emission targets and the constitutional right of citizens to a healthy environment.

At the moment, there is a major expansion and acceleration of mining projects that suggest that new mining developments trump cautionary environmental over-sight in the administration. The current scope of onshore and offshore exploration of oil and gas is unprecedented. While the ANC in power made some concessions by democratising decision-making in the mining sector, it has streamlined and accelerated decision-making in the energy sector proposes in recent years through the fast-tracking methodology, Operation Phakisa. This methodology seeks to enhance planning and sets out concrete aims to boost the economy. The approach chosen economic clusters such as the oceans economy, mining, and agriculture. So far, the methodology has not shown extensive results, but the first steps were already critically discussed looking at state capture of these sectors and question-ing neoliberal tendencies (Engel 2018).

Early arguments in favour of the use of fossil fuels revolved around being a cheap energy input to fuel the gold industry and the grid, its abundance, and its association with modernity and comfort. While coal mining was admitted to have short-term damaging impacts on the environment, long-term damages from min-ing were portrayed as negligible. Fossil-fuel hegemony was constructed with white middle-class consent and suppression of the black majority. Middle-class consent to cheap coal as an input to the grid sustained fossil-fuel hegemony under apartheid. Also, traditional chiefs played a central role in sustaining the MEC.

In theory, EIA provisions under NEMA institutionalised grievance mechanisms and spaces for active civil society engagement. In practice, inclusion in decision-making is often short-circuited and not well suited for the active participation of poor people with low levels of formal knowledge. Efforts to streamline envi-ronmental oversight and shorten periods for the active engagement of citizens suggest a rollback from democratisation efforts in post-apartheid mining affairs. There is a particular emphasis on urgency in the current administration. Accord-ing to decision-makers, urgency is warranted because of the rampant employment crisis, and new fossil-fuel developments promise the creation of new jobs.

The preceding chapters were about the consolidation and expansion of the fossil-fuel sector in South Africa. The next chapters will discuss on how the state and the fossil-fuel industry are challenged by fossil-free social movement mobili-sation. As we will see, activists devise movement tactics and frames to challenge fossil-fuel hegemony in South Africa. In a final chapter, we will also discuss the potentials and limits of international social movement mobilisation against fossil fuels.

Note

1 Under rehabilitation, they understand

> that the land will be returned to a form and productivity in conformity with a land use plan drawn up before mining commences. It implies the establishment of a stable ecological condition that will not deteriorate substantially with the projected land use. It also implies attention to aesthetic considerations so that the mined-out area will not remain conspicuously different from its natural surroundings.
>
> (Chamber of Mines 1981, p. iv)

References

Barnett, C. and Scott, D., 2007. Spaces of opposition: activism and deliberation in post-apartheid environmental politics. *Environment and Planning A*, 39 (11), 2612–2631. doi:10.1068/a39200

BBC, 2018, February 14. South Africa's Zuma crisis: Gupta home raided by police. *BBC*. Available from: www.bbc.co.uk/news/world-africa-43055494

Beinart, W., 2001. *Twentieth-century South Africa*. New York: Oxford University Press, 2nd ed.

Bond, P., 2014. *Elite transition: from apartheid to neoliberalism in South Africa*. London: Pluto Press.

Bond, P. and Friends, 2010. *Climate justice articles*. Available from: http://ccs.ukzn.ac.za/files/bond%20cj%20articles.pdf

Breed, R., 2016, March 18. Big budget cuts for Ezemvelo. *Zululand Observer*. Available from: https://zululandobserver.co.za/104684/big-budget-cuts-for-ezemvelo/

Centre for Environmental Rights, 2013, September 6. *Civil society and community groups call for just administrative action, and appropriate environmental regulation of mining: written submissions to Parliament on aspects of the MPRDA Amendment Bill, 2013*. Available from: https://cer.org.za/news/submissions-civil-society-groupsmprda-amendment-bill-2013

Centre for Environmental Rights, 2014. *Mining and your community: know your environmental rights*. Available from: https://cer.org.za/wp-content/uploads/2014/03/CER-Mining-and-your-Community-Final-web.pdf

Centre for Environmental Rights, 2015, June 17. *Mining companies launch their first attacks on the One Environmental System*. Available from: https://cer.org.za/news/mining-companies-launch-their-first-attacks-on-the-oneenvironmental-system

Chamber of Mines, 1981. *Guidelines for the rehabilitation of land disturbed by surface coal mining in South Africa*. Johannesburg: South Africa.

Chamber of Mines, 1983. *Mining survey 1983*. Johannesburg: South Africa.

City Press, 2014, July 27. Jacob Zuma's family empire. *News 24*. Available from: www.news24.com/Archives/City-Press/Jacob-Zumas-family-empire-20150430

Coaltech, n.d. *About us*. Available from: http://coaltech.co.za/?page_id=33

Cock, J., 2006. Connecting the red, brown and green: the environmental justice movement in South Africa. *In*: R. Ballard, A. Habib, and I. Valodia, eds., *Voices of protest: social movements in post-apartheid South Africa*. Pietermaritzburg: University of KwaZulu Natal Press, 203–224.

Davenport, J., 2013. *Digging deep: a history of mining in South Africa*. Johannesburg: Jonathan Ball.

Department of Environment, n.d. *Operation Phakisa – oceans economy*. Available from: www.environment.gov.za/projectsprogrammes/operationphakisa/oceanseconomy

Eberhard, A., 2011. *The future of South African coal: market, investment and policy challenges* (Working Paper No. 100). Available from: https://pesd.fsi.stanford.edu/publications/the_future_of_south_african_coal_market_investment_and_policy_challenges

Econometrix, 2012. *Economic report: Karoo shale gas development could boost GDP and create hundreds of thousands of jobs* (media release). Available from: www.shell.co.za/about-us/projects-and-sites/the-karoo/econometrix-karoo-shale-gas-report/_jcr_content/par/textimage.stream/1464110534333/d73a28a32064ab15cf3b7c0ebe3cc552e-78b0540e04fe7cb5837a135c148e55f/econometrix-pressrelease.pdf

Engel, U., 2018. The 'blue economy' and operation Phakisa: prospects for an emerging developmental state in South Africa? *In:* J. Schubert, U. Engel and E. Macamo, eds., *Extractive industries and changing state dynamics in Africa: beyond the resource curse.* London: Routledge, 57–74.

Fig, D., 2012. *Fracking and the democratic deficit in South Africa*. Available from: https://hsf.org.za/publications/focus/focus-64/DFig64.pdf

Fine, B. and Rustomjee, Z., 1996. *The political economy of South Africa: from minerals-energy complex to industrialisation*. London: Hurst & Company.

Finkeldey, J., 2020. Lessons from Marikana? South Africa's sub-imperialism and the rise of Blockadia. *In*: C. Grocott and J. Grady, eds., *The continuing imperialism of free trade*. London: Routledge, 113–124.

Fossil Fuel Foundation, 2018. *About*. Available from: www.fossilfuel.co.za/about-us/

Frack Free South Africa, 2016. *Fracking goes to parliament*. Available from: http://frackfreesa.org.za/index.php/2016/12/08/fracking-goes-to-parliament/

Friedman, S., 2006. *Participatory governance and citizen action in post-apartheid South Africa* (Discussion Paper No. 164). Available from: www.ilo.org/wcmsp5/groups/public/-dgreports/-inst/documents/publication/wcms_193613.pdf

General Electric, 2017, September 6. *GE's efficient power generation and air quality control technology is deployed at Kusile Unit One in South Africa*. Available from: https://www.genewsroom.com/press-releases/ge's-efficient-power-generation-andair-quality-control-technology-deployed-kusile

Greenberg, S., 2009. Market liberalisation and continental expansion: the repositioning of Eskom in post-apartheid South Africa. *In*: D. A. McDonald, ed., *Electric capitalism: recolonizing Africa on the power grid*. London: Earthscan, 73–108.

Greenpeace, 2011. *The true cost of coal in South Africa*. Available from: www.greenpeace.org/archive-africa/en/News/news/The-True-Cost-of-Coal/

Greenpeace, 2018, October 29. New satellite data reveals the world's largest air pollution hotspot is Mpumalanga – South Africa. *Greenpeace*. Available from: www.greenpeace.org/africa/en/issues/inspirethemovement/4202/new-satellite-data-reveals-the-worlds-largest-air-pollution-hotspot-is-mpumalanga-south-africa/

Goodman, J., *et al.*, 2020. *Beyond the coal rush: a turning point for global energy and climate policy*. Cambridge: Cambridge University Press.

Hancox, P.J. and Götz, A.E., 2014. South Africa's coalfields – a 2014 perspective. *International Journal of Coal Geology*, 132, 170–254. doi:10.1016/j.coal.2014.06.019

Harper, P., 2015, August 30. King Zwelithini and the 21st century Zulu empire. *City Press*. Available from: https://city-press.news24.com/News/King-Zwelithini-and-the-21st-century-Zulu-empire-20150830

Hedden, S., Moyer, J. and Rettig, J., 2013. *Fracking for shale gas in South Africa: blessing or curse?* (African Future Paper No. 9), 1–12. Available from: www.files.ethz.ch/isn/175033/AF9_6December2013.pdf

Humby, T., 2015. 'One environmental system': aligning the laws on the environmental management of mining in South Africa. *Journal of Energy & Natural Resources Law*, 33 (2), 110–130. doi:10.1080/02646811.2015.1022432

International Consortium of Investigative Journalists, 2016. *Clive Khulubuse Zuma*. Available from: https://offshoreleaks.icij.org/stories/clive-khulubuse-zuma

Khumalo, S., 2018, July 24. Eskom inks R33.4bn loan deal with China Development Bank. *fin24*. Available from: https://www.fin24.com/Economy/Eskom/eskom-inks-r334bnloan-deal-with-china-development-bank-20180724

Leonard, L., 2011. Civil society leadership and industrial risks: environmental justice in Durban, South Africa. *Journal of Asian and African Studies*, 46 (2), 113–129. doi:10.1177/0021909610391049

Leonard, L., 2016. State governance, participation and mining development: lessons learned from Dullstroom, Mpumalanga. *Politikon*, 44 (2), 327–345. doi:10.1080/02589346.2016.1245526

Leonard, L., 2018. Bridging social and environmental risks: the potential for an emerging environmental justice framework in South Africa. *Journal of Contemporary African Studies*, 36 (1), 23–38. doi:10.1080/02589001.2017.1412582

Madonsela, T., 2016. *State of capture report* (Report No. 6 of 2016/17). Public Protector South Africa. Available from: www.saflii.org/images/329756472-State-of-Capture.pdf

Mahlase, M., 2018, July 7. Ramaphosa tells Zulu King Zwelithini that land in Ingonyama trust is safe. *News24*. Available from: www.news24.com/SouthAfrica/News/ramaphosa-tells-zulu-king-zwelithini-that-land-in-ingonyama-trust-is-safe-20180707

Mail & Guardian, 2015, October 7. *Is SA the world's third-best solar location?* Available from: https://mg.co.za/article/2015-10-07-is-sa-the-worlds-third-best-solar-location

Mandela, N., 1995. *Address by President Nelson Mandela at the centennial celebrations of the Hluhluwe-Umfolozi Park and Greater St Lucia Wetland Park, Hluhluwe Game Reserve, KwaZulu-Natal.* Available from: www.mandela.gov.za/mandela_speeches/1995/950430_hluhluwe.htm

Mandela Institute, 2017. *Public regulation and corporate practices in the extractive industry: a South-South advocacy report on community engagement*. Johannesburg: University of the Witwatersrand.

Manson, A. and Mbenga, B. 2014. *Land, chiefs, mining: South Africa's north west province since 1840*. Johannesburg: Wits University Press.

Marinovich, G., 2016. *Murder at small Koppie: the real story of South Africa's Marikana Massacre*. London: Penguin.

Mining Review, 2018, March 15. Cyril Ramaphosa and the challenges facing mining. *Mining*. Available from: www.miningreview.com/cyril-ramaphosa-challenges-mining-south-africa/

Moolman, P., 2017, September 29. 350% increase in a decade: how expensive is electricity in South Africa compared to other countries? *Power Optimal*. Available from: www.poweroptimal.com/350-increase-decade-expensive-electricity-south-africa-compared-countries/

Munik, V., 2010. *The social and environmental consequences of coal mining in South Africa: a case study*. Available from: www.bothends.org/uploaded_files/uploadlibraryitem/1case_study_South_Africa_updated.pdf

Piper, L. and Von Lieres, B., 2016. The limits of participatory democracy and the rise of the informal politics of mediated representation in South Africa. *Journal of Civil Society*, 12 (3), 90–103. doi:10.4324/9781315145341-7

Saul, J.S. and Bond, P., 2014. *South Africa – the present as history: from Mrs Ples to Mandela & Marikana*. Johannesburg: Jacana.

Scholes, R., *et al.*, eds., 2016. *Shale gas development in the central Karoo: a scientific assessment of the opportunities and risks*. Stellenbosch, South Africa: CSIR.

Shubane, K., 1992. Civil society in apartheid and post-apartheid South Africa. *Theoria: A Journal of Social and Political Theory*, 79, 33–41.

Singer, M., 2011. Towards 'a different kind of beauty': responses to coal-based pollution in the Witbank coalfield between 1903 and 1948. *Journal of Southern African Studies*, 37 (2), 281–296. doi:10.1080/03057070.2011.579441

Statistics South Africa, 2013. *General household survey 2012*. Available from: www.statssa.gov.za/publications/P0318/P03182012.pdf

Steyn, P., 2005. The lingering environmental impact of repressive governance: the environmental legacy of the apartheid era for the new South Africa. *Globalizations*, 2 (3), 391–402. doi:10.1080/14747730500367983

Taverner-Smith, R., 1973. *Coal in Natal*. Pietermaritzburg: University of Natal Press.

Vermeulen, J., 2015, May 4. How Eskom handles power crises: 1950s, 70s, and now. *MyBroadband*. Available from: https://mybroadband.co.za/news/energy/124633-how-eskom-handles-power-crises-1950s-70s-and-now.html

Wilkinson, K., 2015, May 14. Did 34% of households have access to electricity in 1994? *Mail & Guardian*. Available from: https://mg.co.za/article/2015-05-14-did-only-32-of-households-have-access-to-electricity-in-1994

World Bank, 2010, April 8. World bank supports South Africa's energy security plans. *The World Bank*. Available from: www.worldbank.org/en/news/press-release/2010/04/08/world-bank-supports-south-africas-energy-security-plans

Part II

Social Movements Fighting Fossil Fuels

4 Counter-Hegemonic Social Movements

To reiterate a key point from the first part of the book, fossil-fuel hegemony is a spatially expansive political project that is integral to any extractivist state. Fossil-fuel hegemony rests on the 'organisation of consent' (Carroll and Ratner 1994, p. 5), which is maintained through elite persuasion, for example through the signifiers 'cheap energy' and 'energy autonomy', on the one hand, and coercion via the securitisation of extraction areas, on the other. There is, thus, a material and a political component to fossil-fuel hegemony. Hegemony is both centralised by the state policy as well as diffused through practices of everyday life and institutions such as companies, universities, or the family home (Carroll and Ratner 1994, p. 6). While the first part of the book was concerned with political actors that entrench fossil-fuel expansion, this chapter analyses the process of fighting fossil fuels through the mobilisation of social movements.

In the following chapter, I first draw attention to important theoretical and methodological considerations that need to be considered looking at social movements protesting against fossil-fuel expansion. This chapter is divided into a theoretical and a methodological part. In the theoretical part, I draw on the spatial aspects of social movement mobilisation and consider the relationship between social movement actors and the spaces they defend. In the second section of the theoretical part, I argue that leaders are integral to the articulation of counter-hegemonic struggles, especially the way struggles are framed. After the two theoretical chapters, I flesh out some methodological considerations concerning qualitative social movement research. I find it particularly important to discuss my own role as a researcher, partly as an observer and, sometimes, as a participant-observer of the movements.

Spaces of Social Movement Struggles

Anti-extractivist struggles are place-based. In some instances, struggles might also spill over and grab the attention of a wider audience concerned about climate change, land degradation, pollution, or indigenous rights. In order to understand the particular role social movements play in creating spaces of resistance, it is useful to understand what 'space' engenders and how it can be (re-)signified by social movement actors. Space is a multifaceted term that is notoriously

DOI: 10.4324/9781003110835-6

difficult to define, and David Harvey (2006, p. 120) called 'any generic definition of space a hopeless task'. An understanding of space encompasses, among other things, distance, measurements, symbolism, and meanings (Massey 1984). However, some scholars, including Harvey himself, have come up with useful dynamic systematisations of the term. Henri Lefebvre (1991, p. 55) thought about space specifically under the conditions of what he calls 'neo-capitalism' where 'more than ever, class struggle is inscribed in space'. According to Lefebvre, space is both *abstract* and *concrete*. Space is *abstract* in that it is embodied in political and economic practices and *concrete* in that it is produced in everyday interactions. People interacting through space acquire certain habits and gain a degree of competency. Representations of space refer to the 'outer' material qualities of space, which congeal the 'mode of production' responsible for their representation.

Representations of space are purposefully created and embodied, for instance, in architecture, art, or monuments. Lefebvre (1991, pp. 38–39) calls these representations the 'dominant space in any society'. Representational space is encoded in the lived realities of space.

Building on Lefebvre, Harvey (2006, p. 131) elaborates that 'spaces and times of representation that envelop and surround us as we go about our daily lives likewise affect both our direct experiences and the way we interpret and understand representations'. Social movement studies attuned to capitalist modes of production can help to balance the demands for analysing the economic structure as well as its contestation (della Porta 2015). Importantly, Nicholls (2007, p. 612) notes that the effects of capitalism are not the same across space but that the 'geographically uneven logic of economic and state power differentiates the grievances that give rise to social movements'. He highlights that mobilisations, for example, can be results of reactions to investments such as mines or disinvestments from social provisions.

While scholars (e.g. Cox and Nilsen 2014; Barker 2013) agree that social movement organisations (SMOs) formulate critiques to bring changes to the political, economic, and social fabric, social movement analysis often lacks geographical contextualisation. Most studies do not go into much detail on how social movements actually re-signify spaces (both materially and verbally) in conflict with their adversaries. In most studies, CONTESTED places such as streets, occupied buildings, or conservation areas just happen to be where mobilisation occurs without much attention being paid to the interactions and meanings given to the place's specificities. According to Miller (2000), if geography is addressed at all in social movement analysis, it is often fraught with two problems. First, the dualism between 'the social' and 'the spatial' is never fully collapsed. The social and the spatial are treated as distinct categories rather than as co-constitutive. Second, spatial units are homogenised to treat spaces as one undifferentiated unit (such as global, national, or regional) (Miller 2000, pp. 4–5). In seminal studies on social movement mobilisation, the specificities of place and space are given less attention than the movements' material resources (e.g. McCarthy and Zald 1977), their repertoires of contention (e.g. Tilly 2006), or their political demands and opportunities (e.g. Kitschelt 1986).

Social movements take a particular shape in the places they act in or try to defend, but they also construct the places around them by finding a language for their surroundings:

> First, the concept of place informs us about why social movements occur where they do and the context within which movement agency interpellates the social structure. Second, the concept of place informs us about the nature of specific movements.
>
> (Routledge [1993] cited in Miller 2000, p. 37)

In Routledge's account of place, he points to social movements' agency to penetrate spaces and the movement's meaning-making process in places. Lefebvre, on the other hand, points to hegemonic productions of space under capitalism. While Routledge looks at mobilisation against hegemonic productions of concrete places, Lefebvre points to the way in which wider socio-spatial relations are hegemonically produced and reproduced. Therefore, Routledge and Lefebvre are looking at two sides of the same coin. Space and place are relational and amorphous concepts that are highly susceptible to the respective context. Some of the literature on social movements have addressed socio-spatial identity formation and the contestation over the meaning of space (Routledge 2017; Nicholls *et al.* 2013; Leitner *et al.* 2008; Miller 2000; Routledge 2000). These scholars emphasise that space can be experienced in practice but relates to broader social relations (abstract space). For example, community members can become aware of a mining project in their neighbourhood. They might find a notification in the local newspaper or see a board announcing a public participation meeting. Mining would certainly have very real impacts on people's everyday lives; it would reconfigure socio-spatial relations, such as through resettlement, visual intrusions, and the creation of new social relations as a result of newly arriving workers in the community. Some community members might find a job in the mine, entering into a wage-relation with the mining corporation, which would alter the way they experience their living space. The notification can be the harbinger of these fundamental changes in the socio-spatial structures of the community.

Spatial analysis of social movements highlights that space is imbued with meanings that are often conflicting. Mobilisation is particularly likely where actors feel a strong collective attachment to a certain place. Once an organised collective gets the impression that a place they value has come under threat, the likelihood of resistance increases insofar as the change conflicts with 'priorities and imaginaries' are held in a given place (Leitner *et al.* 2008, p. 162). Space is not defined by a fixed set of taken-for-granted properties. Rather, activists define spaces creatively by 'cobbling together different spatial imaginaries and strategies' (Leitner *et al.* 2008, p. 158). Spaces are re-articulated and can become re-signified in the process of mobilisation.

In this section, I discussed how space shapes social movement action. Space is encoded with meaning. Social movements use contested spaces and places as a resource for their claim-making. For example, they might occupy a symbolic building or form a human shield around a forest they want to save from logging. The meaning of space is not carved in stone and is subject to political contestation. Extraction companies frame extraction around development and job creation, while anti-extractivist social movements highlight the inherent value of spaces. Under neoliberal extractivism, mining companies and state actors exert considerable pressure on residents living on precious resources to mine. Thus, it is a question of social movement leadership forming counter-hegemonic campaigns to stop extraction.

Thus far, it is clear that social movements are more likely to occur where the people they represent have a close attachment to the place they inhabit, but we cannot simply assume that social movements simply 'pop up' whenever highly valued spaces are under threat. Much like hegemonic alliance building, counter-hegemonic movements need leadership.

Leadership in Social Movement Action

As seen in the section on the Gramscian notion of hegemony, leadership is a central aspect of counter-hegemonic alliance building. Gramsci ascribes a central role to leaders in his conception of hegemony. According to Gramsci (1991, p. 5), intellectual leadership plays a central role in any hegemonic social formation. Indeed, every hegemonic formation produces its own 'strata of intellectuals which give it homogeneity and an awareness of its own function not only in the economic but also in the social and political fields'. Intellectuals, therefore, help to add coherence and self-awareness to a hegemonic order. 'Organic intellectuals' are not marked by special formal qualifications but rather by everyday intellectual work.

> The mode of being the new intellectual can no longer consist in eloquence, which is an exterior and momentary mover of feelings and passions, but in active participation in practical life, as constructor, organizer, 'permanent persuader' and not just a simple orator.
>
> (Gramsci 1991, p. 10)

'Organic intellectuals' are thus different from 'traditional intellectuals' in that the former are consciously inserted into an emerging social formation, while the latter serve the established order.

Leaders play different roles in SMOs. They face up to day-to-day organisational, emotional, and bureaucratic challenges. The topic of leadership in social movements remains quite under-researched. One of the reasons for this might be that the notion of leadership within horizontally oriented social movements is discredited among many participants and is one of the few taboo topics among social movements (Blee 2013, pp. 124–126). Barker *et al.* (2001) suggest that leadership is often associated with domination and is perceived to be incompatible with demands for 'spontaneity'; so, most scholars shy away from touching on this sensitive topic.

Leaders of a social movement represent the movement to the outside and frame its messages. Understanding frames and framing in social movement studies is a key, as frames are the pictures of social reality that are visible to the outside world. By painting these pictures, social movement actors attempt to significantly change social reality. Frame setting is a dynamic process, as movements act in an ever-changing political environment. Therefore, social movements have to update their claim-making in order to stay in touch with the movement base. There are three core functions of frames in social movements (Snow 2013; Johnston and Noakes 2005; Snow and Benford 1988). First, frames help to structure the attention of social issues in terms of what is relevant or irrelevant, 'in frame' or 'out-of-frame'. It offers movement supporters a novel interpretation of social reality and presents the movement's worldview in an alternative way as opposed to business-as-usual (diagnostic framing). Second, frames present solutions to problems as well as strategies on how to get there (prognostic framing). Third, frames have a *transformative function* in that they seek to mobilise people to join a campaign. Usually, the interpretation of events and solutions to problems is not enough to mobilise people. Therefore, frames need to include a call to action (motivational framing). Frames are thus able to transform 'routine grievances or misfortunes into injustices or mobilising grievances' (Snow 2013, p. 1). Frames are more than the sum of individually held beliefs of movement supporters; they are the purposeful outcome of negotiations over meaning (Snow and Benford 2000, p. 614). Frames have both internal and external signalling functions. To movement adherents, frames clarify and consolidate a movement's political position. To target audiences, frames render the movement recognisable and motivate sympathisers to join.

Frames are crafted by engaged people. Star anthropologist and recently deceased social justice activist David Graeber is credited with crafting the slogan 'We are the 99%', which continues to be widely used in the Occupy movement and beyond (Szolucha 2017). Frames are purposefully used by movement leaders to mobilise people and communicate their ideas to the public. Leaders in social movements are 'strategic decision-makers who inspire and organise others to participate in social movements' (Morris and Staggenborg 2004, p. 171). Leaders are also the main actors in processes of frame articulation and amplification (Johnston and Noakes and 2005, p. 8). Leaders often use the cultural stock that surrounds them to make their claims rather than coming up with something entirely new. Other frames are intentionally borrowed from relevant cultural scripts in a movement's environment. Their messages get amplified when a movement's symbols and slogans proliferate in different political contexts (ibid.). Sometimes, a particularly powerful master signifier operates as the political horizon of a movement. For a movement's success, the 'flexibility and inclusivity' of master frames are key (Benford and Snow 2000, p. 619).

Framing processes within social movements can be contentious. In fact, most movements aspire to be horizontal so that everyone can equally be a part of the campaign framing. In reality, however, most movements are hierarchical. Some movements have designated spokespeople, and in such cases, few movement activists take part in drafting press releases and postings on websites and social media (Gerbaudo 2017). Spokespeople do not communicate the movement's framing in a vacuum. A successful framing is undertaken in close contact with rank-and-file members and in accordance with social codes. Otherwise, movements risk demobilisation.

However, leaders of smaller social movement groups 'are often torn between institutionally embedded demands for organisational efficiency and culturally embedded values for inclusive engagement' (Choi-Fitzpatrick 2015, p. 123). Often, there is an immanent tension between professed claims to leaderless decision-making and the reality of (unacknowledged) leadership. Nunes (2014, p. 13) proposes that leadership is a necessary element of organisation and criticises the fantasy of 'absolute horizontalism'. Rather, a sweet spot 'between openness and closure, dispersion and unity, strategic action and process' needs to be negotiated by the members (Nunes 2014, p. 13). 'Creative' forms of leadership are 'multidimensional and frequently *distributed*' (Krinsky and Crossley 2014, p. 6, emphasis in original).

> [A] leadership group's dense connections to grassroot[s] activists on the one hand . . . [and] leaders' ability to react creatively to [a] situation depends on strong local knowledge and motivation as well as the ability to recontextualize this knowledge through dialogue with experiences gained in other movement settings.
>
> (Krinsky and Crossley 2014, pp. 5–6)

The study of leadership, therefore, invites scholars to look beyond particular leadership figures and understand their social embeddedness in the groups they are a part of (Sutherland *et al.* 2014; Alvesson and Spicer 2012). Studying leadership involves more than simply identifying the attributes of certain key figures and indeed attunes scholars to look at the functions and role of leaders in relation to the group to which they belong.

Leaders are also responsible for opening up channels to connect to other movements or causes. Some work has also gone into thinking about the composition of social movement alliances. Krinsky and Reese (2006) look into alliance-building and collaboration between activist groups. The collaboration between unions and social movements, for example, is a possible win–win scenario with social movement tactics enriching the repertoires of contention of the unions and the movements gaining from the unions' resources. However, Krinsky and Reese (2006, p. 627) identify that limitations in collaborations between unions and community-based organisations (CBOs) are particularly pertinent in the US, where working-class issues are historically deeply entrenched among white males. Different organisations making similar claims might conflict over attention or resources, triggering competition rather than collaboration (Krinsky and Reese 2006, p. 630). Other fault lines standing in the way of alliance-building might be class, geography, and culture (Beamish and Luebbers 2009, p. 648). This can be mitigated by the presence of 'bridge-builders' who have 'overlapping identities, experiences, or movement ties' (Krinsky and Reese 2006, p. 650). Krinsky and Reese (2006) and Isaac and Christiansen (2002) show that the respective historical and institutional contexts in which mobilisation takes place matter. Some periods offer greater opportunities for building alliances. For example, heightened threats or opportunities are conducive to a movement's convergence (Beamish and Luebbers 2009, p. 650).

An Ethnographic Approach to Social Movements

This research takes a qualitative approach to social movement mobilisation. Quantitative scholars use datasets and computational models to test theories or probabilities for movements' rise and fall, among other operations. While these studies are impressive in their scope and cross-country observations, they mostly lack historical and geographic sensitivities. Additionally, they lack the self-interpretations and meaning-making of activists themselves. Qualitative methods in social movement studies render visible the context-specific and historical patterns of specific social movements. Most importantly, qualitative social movement scholars immerse themselves in the field trying to understand the history and meaning-making of particular movements. In the process, they talk to a number of activists, follow their meetings, and attend their protests. I adopted this latter approach.

In this book, I look at social movements standing up to the fossil industry that has strong ties to entrenched interests at the state level. I undertook an ethnographic study to understand the inner life of fossil-free movements in South Africa in the spirit of 'getting out of the armchair' (Balsiger and Lambelet 2014). During my fieldwork, I used selected qualitative methods, including expert interviews, participant observations, and focus groups, among others. The main focus of my study was on activists practices, but I also interviewed politicians, corporate actors, journalists, and people working in the concerned ministries. This informed my understanding of the interests and practices culminating in fossil-fuel hegemony.

I carried out two periods of fieldwork in South Africa: January to March 2016 and September 2016 to March 2017. I undertook in-depth research on the communities and activist organisations opposing fossil-fuel projects with a focus on the KwaZulu-Natal (KZN) province. During my fieldwork, I was based at the Centre for Civil Society (CCS) at the University of KwaZulu-Natal, which constituted my institutional home away from home (Kern and Vossiek 2014). The Centre has made a name around Africa and beyond for its critical and engaged bottom-up research. At the Centre, I was a part of debates concerning Africa's resource-cursed economies and social movement practices. The discussions were particularly helpful to understand the grievances and needs of civil society actors in South Africa and some other African countries. The Centre also offered a space for critical self-reflection on what it means to be a white European researcher in a postcolonial country who is studying resource conflicts involving historically disadvantaged South Africans.

Prior to my research on the South African mining sector, I conducted research on and visited the Marikana area where a miners' strike was gunned down by South African police forces in 2012. In the following years, these events received wide scholarly attention. Marikana is a platinum mining area, and it is mined by multinational companies that sell the product on the world market. The Marikana massacre brought to light the ongoing violent undercurrents in the South African mining sector. It made me start to take a closer look at the dynamics around contentious resource frontiers in South Africa and their global implications that led me to take a more ethnographic approach. The Marikana massacre also marked a new beginning of more critical scholarly engagement with the South African mining sector that I draw on in my analysis.

Reflexivity and Participant Research Practices

As my fieldwork on anti-fossil-fuel movements proceeded, my position shifted from observer to participant observer. At the beginning of my research on each campaign, it was hard to convey what my role as a researcher would actually involve. Most of the activists were happy to meet for an interview, but some of them did not quite grasp that my research would be conducted over several months and would involve participation throughout that time. There were also situations where the line between the role of a researcher and an activist were blurred. The following excerpt from my research diary reflects this:

> [At] night, we were busy preparing for the march. I found myself writing, 'Leave the coal in the hole' or '#CoalMineMustFall' on the banners. Removing the paint on my fingers under the shower that night reminded me that I had become something more than just a bystander.
>
> (Field notes 24 March 2016)

On the day of the march mentioned in this diary entry, participants voiced their appreciation and thanked me for having participated in the protest. I said that I had merely taken pictures and observed the march rather than joining in. To my surprise, I was told that participants would have otherwise expected police violence or fights between rival protesters. When I asked for an explanation, I was told that protesters had anticipated violence in the absence of a handful of white journalists, NGO activists, and scholars. According to some participants, it was because of 'you white people' that violence had not erupted that day.

This discussion after the protest march showed me that it would be disingenuous to pretend that I was a detached observer disinterestedly documenting social movement practices and that even as a witness, I had played an active part. I believe that this realisation is necessary to gain a good understanding of what is actually at stake when discussing social movement action. I have great sympathy for the movements' causes. While some might call this bias, 'neutrality' in highly politicised settings such as social movement campaigns is an illusory concept. Rather, it should be acknowledged that the researcher's personal opinions and values do influence the choices taken during the research process.

Interviews

The 28 formal interviews I conducted and countless informal discussions I had with members of the SAVE and Frack Free campaigns are particularly valuable, as the mobilisations had not been researched previously. Although both groups maintain well-updated web pages, these do not address, for example, the questions around internal processes concerning the organisations and their tactics. Once I became acquainted with some people campaigning for SAVE or Frack Free, it was relatively easy to recruit participants for my study using snowball sampling. In the absence of formal membership lists, snowball sampling is particularly useful (Blee 2013, p. 604).

My role and stance in these interviews varied. As McDowell (1998, p. 2138) emphasises, researchers can adopt different roles to suit different interview situations. In her study of elites in the City of London, she highlights how she played 'dumb' with an 'older charming but rather patriarchal figure' but was 'sisterly' with similarly aged women (McDowell 1998, p. 2138). Most of my interviewees were older than me, so I adopted the role of being a respectful learner, which was not dissimilar to the position a pupil takes in relation to teachers. In the case of the high-status elites, I was especially careful not to challenge the interviewees' authority and was rather passive about waiting for my turn to speak. I tried to be subtle and patient while steering them to the subjects I wanted to address. Interviews with younger professionals who were roughly my age or slightly older had a more casual and relaxed tone; they were more interactive and felt less hierarchical. In some interviews, I had to encourage participants. Right at the beginning of one of my most informative interviews, the interviewee said somewhat anxiously, 'I don't know what I can say' (interview with activist, 11 January 2017). Another interviewee was concerned that his command of the English language would be inadequate. In both instances, I reassured the participants that their contribution was valuable to my research project, and in order to make them feel comfortable, I tried to avoid any controversial questions early on in the interview.

In-depth interviews are a 'fundamental tool for generating empirical knowledge' (della Porta 2014, p. 228), and they help the researcher to make sense of detailed social processes within social movements as well as the motivations of individuals (Blee 2013, p. 603). Activist interviews help to provide an overview of the rich spectrum of activists and offer insights from rank-and-file members who have a rich historical memory of the movement in which they are active. They 'are a key source of information . . . including how individual and collective identities are constructed, how movements attract followers, and how members' commitments fluctuate over time' (Blee 2013, p. 603). As the literature emphasises, semi-structured interviews are particularly valuable for understanding how groups organise internally by getting behind the scenes as well as for the study of under-researched, 'loosely organized . . . [and] thinly documented social movements' (Blee and Taylor 2002, p. 93). Insider stories help researchers to see the struggles through activists' eyes, which can provide insights into the emotional bonds between activists (Jasper 2014; Blee and Taylor 2002, p. 96).

During the interviews, I would sometimes alter the order of the questions or add some check back questions, while making sure that all relevant questions were answered. I played the role of an understanding ally and used an amicable conversational tone. Johnson even goes as far as to liken in-depth interviews to the relation one has to close friends because interview questions touch on 'personal matters, such as an individual's self, lived experience, values and decisions, occupational ideology, cultural knowledge, or perspective' (2001, p. 104). I also made sure to leave plenty of time for clarifications, follow-up questions, longer elaborations, and even digressions to reveal useful information. In the early stages of research, I left more room for open-ended questions, whereas towards the end, I focused more on filling gaps (Blee and Taylor 2002, pp. 99–100). Generally, interviewees seemed pleased to have the opportunity to reflect on their experiences as activists, even if the interviews sometimes caused inconvenience. One rural activist (interview with activist, 27 March 2016), for example, said that she really welcomed the opportunity to be interviewed, although whenever she had a white visitor, her neighbours thought that she was receiving payment and would almost certainly call by for a share afterwards!

There are factors that make every interview situation unique, and interviews are influenced by external factors such as place and time. Interviews with middle-class activists were typically undertaken in cafés. In most cases, the interviewee would suggest a café of their choice where we would then meet. This guaranteed that the interviewees felt comfortable, and I often started the interview by asking about the meaning of the place for the interviewee. Sometimes, the location of a café provided me with insights into the interviewee's motivations for becoming an activist. During one interview at a café overlooking a small river on the outskirts of Pietermaritzburg, my interviewee kept being distracted by birds she spotted from the terrace where we sat. She would often interrupt her account of Frack Free campaigning to point out particular birds to me. At the time, I was worried that we would not be able to touch on all the points I wanted to cover in a systematic way. Later, I realised that this expression of her love for birdlife and conservation illustrated her motivation to campaign against fracking, which she believed would destroy the habitat of birds and other wildlife.

Despite having relatively easy access to the members of both campaigns, some puzzles remained. It took some time for me to understand how they were organised, the interlinkages between them, and the different roles of the organisations involved. These puzzles reveal the limitations of interviews and the value of participant observation in understanding loosely organised social movements.

Observation and Participant Observation

Throughout my fieldwork, I made observations that were relevant to an understanding of the politics and complex decision-making around mining in KZN. This required a lot of flexibility, because most events or meetings were not announced in advance. Flexibility in qualitative research is seen as a 'necessary tool – not as a concession or failure – and as a tool that can be used to the researcher's advantage' (Billo and Hiemstra 2013, p. 317). Remaining flexible helps the researcher allow for 'revision, evolution, and feedback' (Billo and Hiemstra 2013, p. 317). Observations also help to get an impression of the interactions between the participants of the study, particularly for understanding the networks among social movement campaigners.

For the observations, I relied on gatekeepers to agree to my presence. It was a key to establishing trust and rapport with the leaders of the activist groups. Luckily, I was able to build good personal relationships with activists from both the SAVE and FFSA campaigns. For example, I was invited to join the FFSA WhatsApp group, which helped me to stay updated on information regarding events. After I gained the trust of the activists, I was invited to join them for an array of different events, including social events such as informal dinner parties. Conversations at these social occasions enabled me to find out about upcoming activities and understand the inner-life of these movements.

Data Analysis and Triangulation

Making use of several methods, document types, theories, and viewpoints can increase the accuracy and reliability of social movement research (Ayoub *et al.* 2014, p. 67; Corbin and Strauss 2008, p. 28). Thus, triangulation was an important part of my data analysis. Despite some scepticism about the eclectic nature of triangulation, there is agreement that it has its merits in terms of 'justifying and underpinning knowledge' (Flick 2004, p. 169). Arguably, it 'allows for the analyst to paint a more holistic picture of the complex phenomenon that social movement scholars study' (Ayoub *et al.* 2014, p. 68).

Throughout my data analysis, I was continually triangulating diary entries and recording my observations of events, actions, activities, interview transcripts, and documentary sources. Triangulation was particularly useful for controlling for unavoidable biases of activists in favour of the movements they were part of (Blee 2013, p. 603). I employed certain methods to control for inaccuracies because of bias, such as comparing independent media sources and transcripts from interviews with activists on the same questions.

While some scholars wait until the completion of fieldwork before starting data analysis (Mattoni 2014), I started preliminary processing and analysis while still in the data collection stage of my research. Whenever possible, I tried to transcribe interviews soon after they took place to triangulate with other sources and make time for reflection. However, it was a balancing act to navigate between the demands of collecting enough information and thinking through the ways in which I could adapt and narrow my research focus. On the one hand, there is a need to remain reasonably open to incorporating new insights emerging from the field. On the other hand, there is a danger of broadening rather than narrowing the focus during fieldwork. Therefore, it was important to keep track of my research activities and reflect on ways to improve or alter my strategies. The key to my thought process was that I kept a research diary where I recorded my ideas and observations. Re-reading my own thinking about the research helped me to develop and refine my understanding of the processes I was studying. As Bryant and Charmaz (2007, p. 1) state, this 'back and forth between empirical data and emerging analysis makes the collection of data progressively more focused'.

Codes were used to structure the large amounts of text from my field notes and the interview material. Finding themes helps the researcher to move beyond the chronological order of events and to identify 'topics that are reoccurring, uniquely illustrative, and/or otherwise prominent' (Harrison 2018, p. 126). Reflecting on coding, Miles and Huberman (1994, p. 55) emphasise how researchers need to reduce complexity in a 'selective process'. Codes are described as 'tags or labels for assigning units of meaning to the descriptive . . . compiled during the study' (Miles and Huberman 1994, p. 56). This is not to discredit the experimental and intuitive process of organising themes but to emphasise the fact that intuition needs to be combined with systematic rigour (Harrison 2018, pp. 126–127). I came up with preliminary codes before my field trips based on my research questions and main themes. After my first field trip, I reviewed the codes by merging some and deleting others. Field notes proved to be much more useful for deriving codes than I had anticipated when I was first taking the notes (often forcing myself to write them directly after long hours in the field).

There were different coding stages during my research. At first, I assigned broader codes to the transcribed interviews. These codes were meant to discern the relevant chunks in the responses. Subsequently, I came up mostly with descriptive codes but also included some early interpretations. From my early research questions, sub-questions, and problem areas, I deduced codes to assign to the transcripts, which I kept on a separate sheet of paper. For example, these preliminary codes included 'environmental assessment', 'economic case for mining', 'geography', 'resistance', and 'leadership' (all of these codes were retained). The same was done with the field notes. To test the validity of the codes, I then compared the codes assigned to different field notes. Codes that were recurring were retained for the next stage of coding, and others were deleted since they proved to be less relevant. During my field research, I wrote additional field memos that were very useful for improving my codes. For the second stage of coding, I merged some of the codes and aimed to make them more 'conceptually inclusive' (Miles and Huberman 1994, p. 62). According to Miles and Huberman (1994, p. 62), the high-quality codes, taken together, follow a 'governing structure' that ensures coherence between the parts of the analysis. To improve the codes and clarify their meaning, I remained in close contact with colleagues and my supervisors.

In March 2017, i.e. towards the end of my second period of fieldwork, I reached a point where I felt during the interviews that the answers resembled responses from previous interviews. Issues that were raised sounded familiar and added little to my understanding of the two campaigns' collective stories despite obvious differences in activists' personal accounts of their engagement.

I realised that I had reached a saturation point: as Johnson (2001, p. 114) emphasises, when the feeling that 'the learning curve has peaked' sets in, it may be because of saturation. Although there is no agreement about how many interviews are sufficient to reach saturation, I had conducted over 40, which exceeds the numbers discussed in relevant publications. After reaching saturation point, I focused more on disseminating and discussing my preliminary findings among the interested public.

Being able to present my findings to different audiences in South Africa was very helpful for making sense of the processes I had documented and observed. I presented my findings and ideas at two academic conferences in Johannesburg, at two research seminars at the University of KwaZulu-Natal, at two NGOs and to the government conservation organisation Ezemvelo Wildlife. Through debates with academics, professionals and activists, I had the chance to challenge my thinking and revise my arguments. My seminar with Frack Free activists was particularly instructive. It reinforced just how contested fossil-fuel extraction is in KZN and underscored the political character of fossil-fuel expansion. Activists invited me to speak and decided on the title 'How to fight the extractive industry' for the seminar. I was fine with the title, but I was slightly uncomfortable with the idea of giving advice to activists in my presentation.

The talk was rather factual, which disappointed some of the activists present. However, the seminar was also attended by an environmental consultant employed by Rhino Oil and Gas, the prospecting company that is pursuing applications for fracking in the Midlands and Newcastle. The consultant obviously considered it important to attend the activist discussion to gauge the level of discontent and the issues at stake within the movement. His presence caused some unease among FFSA activists, who felt they were being observed by an associate of their opponents. After my presentation, the consultant told me that he wondered if the activists present ever thought about the consequences of their demand to stop fossil-fuel expansion. He insisted that activists relied on fossil fuels as much as anyone else for transport and household energy and called them hypocrites for not acknowledging this.

When I returned from South Africa after my second period of fieldwork, I spent six months transcribing and processing my data. In the coding phase, I was able to identify the re-occurring themes and obtained a broader understanding of the underlying processes. After the coding phase, I re-read the interviews and documents I had collected and drafted the empirical chapters and discussion. During this process, I went back and forth between my theoretical framework and the empirical findings. After spending so much time in the field, I felt more like an investigative journalist than an academic. I had collected rich data but found it easier to write investigative articles and current analyses of fossil-fuel extractivism. I found it hard to find a way back into making an academic contribution and finding a theoretical approach. It was at this point that I did more extensive reading on extractivism and climate change as well as debates around counter-hegemonic social movements in order to frame this study.

References

Alvesson, M. and Spicer, A., 2012. Critical leadership studies: the case for critical performativity. *Human Relations*, 65 (3), 367–390. doi:10.1177/0018726711430555

Ayoub, P., Wallace, S. and Zepeda-Millán, C., 2014. Triangulation in social movement research. *In*: D. Della Porta, ed., *Methodological practices in social movement research*. New York, NY: Oxford University Press, 67–96.

Barker, C., Johnson, A. and Lavalette, M., eds., 2001. *Leadership and social movements*. Manchester, England: Manchester University Press.

Barker, N., 2013. What would a Marxist theory of social movements look like? *In*: C. Barker, L. Cox, J. Krinsky and A.G. Nilsen, eds., *Marxism and social movements*. Leiden, Netherlands: Brill, 63–82.

Balsiger, P. and Lambelet, A., 2014. Participant observation. *In*: D. Della Porta, ed., *Methodological practices in social movement research*. New York, NY: Oxford University Press, 144–172.

Beamish, T.D. and Luebbers, A.J., 2009. Alliance building across social movements: bridging difference in a peace and justice coalition. *Social Problems*, 56 (4), 647–676. doi:10.1525/sp.2009.56.4.647

Benford, R.D. and Snow, D.A., 2000. Framing processes and social movements: an overview and assessment. *Annual Review of Sociology*, 26, 611–639.

Billo, E. and Hiemstra, N., 2013. Mediating messiness: expanding ideas of flexibility, reflexivity, and embodiment in fieldwork. *Gender, Place & Culture*, 20 (3), 313–328. doi:10.1080/0966369x.2012.674929

Blee, K.M., 2013. *Democracy in the making: how activist groups form*. New York, NY: Oxford University Press.

Blee, K.M. and Taylor, V., 2002. Semi-structured interviewing in social movement research. *In:* B. Klandermans and S. Staggenborg, eds., *Methods of social movement research*. Minneapolis, MN: University of Minnesota Press, 92–117.

Bryant, A. and Charmaz, K., 2007. *The Sage handbook of grounded theory*. Thousand Oaks, CA: Sage.

Carroll, W.K. and Ratner, R.S., 1994. Between Leninism and radical pluralism: Gramscian reflections on counter-hegemony and the new social movements. *Critical Sociology*, 20 (2), 3–26.

Choi-Fitzpatrick, A., 2015. Managing democracy in social movement organizations. *Social Movement Studies [Online]*, 14 (2), 123–141. doi:10.1080/14742837.2014.945158

Corbin, J.M. and Strauss, A.L., 2008. *Basics of qualitative research: techniques and procedures for developing grounded theory*. Thousand Oaks, CA: Sage, 3rd ed.

Cox, L. and Nilsen, A.G., 2014. *We make our own history: Marxism, social movements and the crisis of neoliberalism*. London: Pluto Press.

Della Porta, D., 2014. In-depth interviews. *In: Methodological practices in social movement research*. New York, NY: Oxford University Press, 228–261.

Della Porta, D., 2015. *Social movements in times of austerity: bringing capitalism back into protest analysis*. Cambridge: Polity.

Flick, U., 2004. Triangulation in qualitative research. *In*: U. Flick, E. Von Kardoff and I. Steinke, eds., *A companion to qualitative research*. Thousand Oaks, CA: Sage, 178–183.

Gerbaudo, P., 2017. Social media teams as digital vanguards: the question of leadership in the management of key Facebook and Twitter accounts of Occupy Wall Street, Indignados and UK Uncut. *Information, Communication & Society*, 20 (2), 185–202.

Gramsci, A., 1991. *Selections from the prison notebooks*. London, England: Lawrence and Wishart.

Harrison, A.K., 2018. *Ethnography*. New York, NY: Oxford University Press.

Harvey, D., 2006. *Spaces of global capitalism: towards a theory of uneven geographical development*. Brooklyn, NY: Verso.

Isaac, L. and Christiansen, L., 2002. How the civil rights movement revitalized labor militancy. *American Sociological Review*, 67 (5), 722–746. doi:10.2307/3088915

Jasper, J.M., 2014. *Protest: a cultural introduction to social movements*. Cambridge: Polity.

Johnson, J.M., 2001. In-depth interviewing. *In*: J.F. Gubrium and J.A. Holstein, eds., *Handbook of interview research: context and method*. Thousand Oaks, CA: Sage, 103–119.

Johnston, H. and Noakes, A.N., eds., 2005. *Frames of protest: social movements and the framing perspective*. Lanham: Rowman & Littlefield.

Kern, F.G. and Vossiek, J., 2014. Get organised: The 'do's' preceding successful field research. *European Political Science*, 14 (2), 137–148. doi:10.1057/eps.2014.45

Kitschelt, H.P., 1986. Political opportunity structures and political protest: anti-nuclear movements in four democracies. *British Journal of Political Science*, 16 (1), 57–85. doi:10.1017/s000712340000380x

Krinsky, J. and Crossley, N., 2014. Social movements and social networks: introduction. *Social Movement Studies*, 13 (1), 1–21. doi:10.1080/14742837.2013.862787

Krinsky, J. and Reese, E., 2006. Forging and sustaining labor – community coalitions: the workfare justice movement in three cities. *Sociological Forum*, 21 (4), 623–658. doi:10.1007/s11206-006-9036-0

Lefebvre, H., 1991. *The production of space*. London, England: Verso.

Leitner, H., Sheppard, E. and Sziarto, K.M., 2008. The spatialities of contentious politics. *Transactions of the Institute of British Geographers*, 33 (2), 157–172. doi:10.1111/j.1475-5661.2008.00293.x

Massey, D., 1984. Geography matters! *In*: D. Massey and J. Allen, eds., *geography matters!: a reader*. Cambridge, MA: Cambridge University Press, 1–11.

Mattoni, A., 2014. The potentials of grounded theory in the study of social movements. *In*: D. Della Porta, ed., *Methodological practices in social movement research*. New York, NY: Oxford University Press, 21–42.

McCarthy, J.D. and Zald, M.N., 1977. Resource mobilization and social movements: A partial theory. *American Journal of Sociology*, 82 (6), 1212–1241. doi:10.1086/226464

McDowell, L., 1998. Elites in the city of London: some methodological considerations. *Environment and Planning A*, 30 (12), 2133–2146. doi:10.1068/a302133

Miles, M.B. and Huberman, A.M., 1994. *Qualitative data analysis*. Thousand Oaks, CA: Sage, 2nd ed.

Miller, B.A., 2000. *Geography and social movements: comparing antinuclear activism in the Boston area*. Minneapolis, MN: University of Minnesota Press.

Morris, A.D. and Staggenborg, S., 2004. Leadership in social movements. *In*: D.A. Snow, S.A. Soule and H. Kriesi, eds., *The Blackwell companion to social movements*. Malden, MA: Blackwell Publishing, 171–196.

Nicholls, W.J., 2007. The geographies of social movements. *Geography Compass*, 1 (3), 607–622. doi:10.1111/j.1749-8198.2007.00014.x

Nicholls, W.J., Miller, B. and Beaumont, J., eds., 2013. *Spaces of contention: spatialities and social movements*. London: Routledge.

Nunes, R., 2014. *Organisation of the organisationless: the question of organisation after networks*. Lüneburg: Post-Media Lab.

Routledge, P., 1993. *Terrains of resistance. Nonviolent social movements and the contestation of place in India.* Westport, CT: Praeger.

Routledge, P., 2000. Geopoetics of resistance: India's Baliapal movement. *Alternatives: Global, Local, Political*, 25 (3), 375–389. doi:10.1177/030437540002500312

Routledge, P., 2017. *Radical geographies of protest.* London: Pluto Press.

Snow, D.A., 2013. Framing and social movements. *In:* D.A. Snow, D. Della Porta, B. Klandermans and D. McAdam, eds., *The Wiley-Blackwell encyclopedia of social and political movements.* Hoboken, NJ: Blackwell.

Snow, D.A. and Benford, R., 1988. Ideology, frame resonance and participant mobilization. *International Social Movement Research*, 1, 197–217.

Snow, D.A. and Benford, R., 2000. Framing processes and social movements: an overview and assessment. *Annual Review of Sociology*, 26, 611–639.

Sutherland, N., Land, C. and Böhm, S., 2014. Anti-leaders(hip) in social movement organizations: the case of autonomous grassroots groups. *Organization*, 21 (6), 759–781. doi:10.1177/1350508413480254

Szolucha, A., 2017. *Real democracy in the occupy movement: no stable ground.* London and New York: Routledge.

Tilly, C., 2006. *Regimes and repertoires.* Chicago, IL: University of Chicago Press.

5 Leadership and Framing in Fuleni's Anti-Coal Movement

The Environmental Justice Atlas (EJA) lists 322 conflicts over coal infrastructure globally until May 2022, including planned mines or coal-fired power stations (EJA n.d.). Conflicts often involve local residents, indigenous groups, and NGOs organising social movement actions together or individually. In recent years, global environmental NGOs focused their work on resistance to coal, emphasising the contributions of coal burning to climate change (e.g. Climate Action Network Europe 2021; Greenpeace 2008). Resistances to coal can be found in all major coal-producing countries, including China, India, the United States, Indonesia, Australia, Russia, Germany, and South Africa. Some countries in the Global North have tabled coal phase-out strategies (e.g. German Institute for Economic Research *et al*. 2019), whereas developing countries are building up coal capacity and are often supported by Western donors. Gellert and Ciccantell (2020, p. 216) therefore caution not to prematurely proclaim the end of global coal: 'even when core Western states reduce their coal extraction and consumption, transnational companies based in these states continue to engage in coal mining investment and accumulation'.

Movements' repertoires of contention include lawsuits, marches, occupations, alternative information dissemination campaigns, and creative performances to draw attention to the negative impacts of coal. The demands range from shelving plans for the construction of coal mines or coal-fired power stations to receiving fair compensation and proper public participation. Levels of conflict around coal are typically quite high and, sometimes, involve acts of physical violence and even the death of activists. Complaints are raised at all stages of coal infrastructure development, including construction and operation (Environmental Justice Atlas n.d.).

As shown in Chapter 3, coal is particularly entrenched in South Africa as a major fossil-fuel developing country. The construction of two of the world's largest coal-fired power stations and the opening of new mines ensure coal dependence in South Africa for the next decades. Fuleni Mine project is a particularly lucrative project, since it looks at high-grade coal development. The pressure to open new coal mines in South Africa remains high. As discussed earlier, mining projects are inserted into existing political, economic, and social contexts. The particular context needs to be taken into consideration when we look at mine conflicts involving

DOI: 10.4324/9781003110835-7

social movements. In the following, I will discuss the place history and contested land uses in the area.

Place History and Land Use in Fuleni

Historically, Fuleni is a contested space where many battles were fought prior to the current mining conflict. As we will see, the history of the Hluhluwe–iMfolozi Park plays an important role in the way current conflicts around the area are framed. Today, nature conservation has very different consequences and implications for the local residents as opposed to foreign visitors of the conservation area.

The Zulu Kingdom, believed to be the most centralised form of African governance in the 19th century (Mamdani 1996, p. 44), was founded by King Shaka (1816–1828). It lasted until the end of the rule of his nephew Cetshwayo (1872–1879). The core of the Zulu Kingdom encompassed five rivers in today's KZN, with the Umfolozi River at its centre. The area covered by the Zulu Kingdom was a 'fertile, well-watered region, ideal for cattle-raising, largely free of debilitating diseases and protected by a formidable military organization' (Guy 1982, p. 14). Favourable geography and military might have translated into unparalleled independence from colonial invaders in Southern Africa. Attempts to incorporate the Zulu population into the waged labour force by colonial forces were largely unsuccessful. Additionally, foreign trade and Christian missions were not able to make inroads in the area (Guy 1982, p. 15).

The Zulu victory at Isandlwana against British forces in 1879 was perceived as a 'stain' on Britain's military record. The Zulus' independence only began waning later that year with a major defeat against the British at Ulundi, which involved heavy losses (Guy 1982, pp. xx–xxi). Fragmentation ensued, and substantive alterations to everyday life penetrated the crumbling Zulu Kingdom. For example, the imposition of the hut tax would not have been possible without the military defeat by the British (Guy 1982, p. 231).

Not long after, the first nature reserves in Africa were declared in Hluhluwe and iMfolozi in 1895. Along with the US, South Africa became a trailblazer in nature and game conservation. From the 1920s, more and more white South Africans were seeking an escape from industrialised lives in the cities (Brooks 2005, p. 223). Conserving nature was also a means to unite English and Boer nationalists. Imfolozi was particularly interesting, as it housed the white rhinoceros, which enjoyed particular protection. Accommodation for overnighting tourists was built as early as 1937 in Hluhluwe.

Conflicts over land use erupted in the mid-1930s when the Wild Life Protection and Conservation Society clashed with the Zululand Farmers Union over future land use. While the former advocated the fusion between the two nature reserves, the Zululand farmers voiced their opposition to this project. The land in between the two reserves, which is referred to as the 'Corridor' (see Figure 5.1), was finally incorporated into the park. Zulu people living in the Corridor were not considered in decision-making according to the records available today. Forced removal

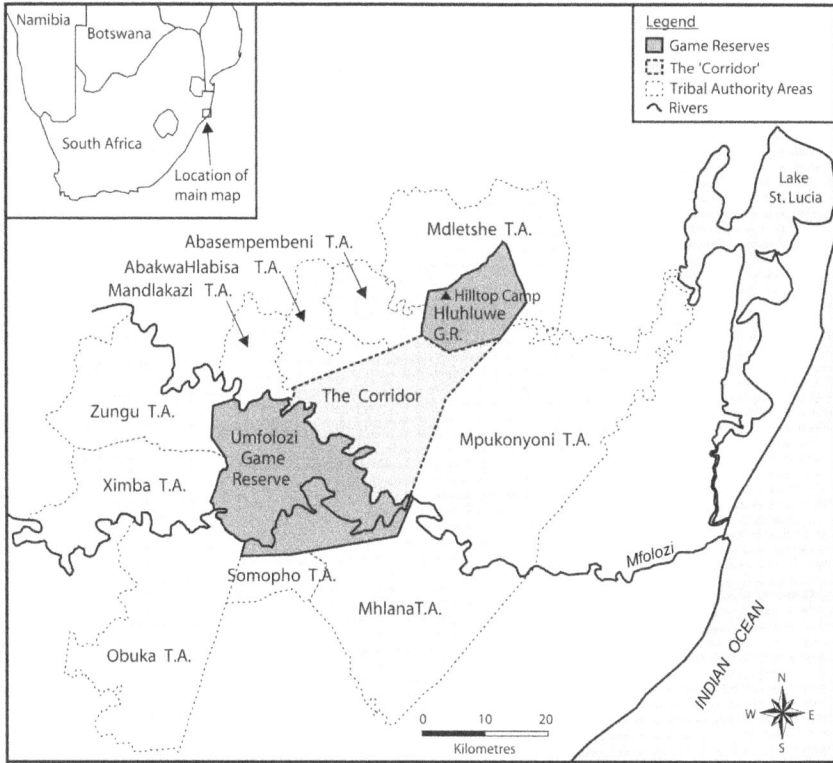

Figure 5.1 Map of Hluhluwe–iMfolozi Park and surrounding area; mass evictions took place in the 1930s to declare the Corridor a conservation space.

Source: Brooks (2005)

started in the 1940s. Brooks (2005) describes the land-use practices around the area as part of a 'brutal geography' making.

From its inception, black game guards and guides were working in the park, and according to historical records, they were reasonably content with the work they were given (Brooks 2005, p. 229). White rangers learnt about Zulu culture and tradition for their jobs in the park (Brooks 2008, p. 293). Underneath what appeared to be harmonious relations, the park meant different things to Zulus and white English-speaking rangers. In the English language, the park was synonymously referred to as 'Shaka's game reserve' or 'Shaka's royal hunting ground'. In Zulu, the reserve was equally referred to as 'indawo yenkosi yokuzingela', which translates to 'royal hunting grounds'. However, the area was also referred to as 'isiqiwu', which translates to 'beacon' as well as 'boundary maker'. 'Beacon' referred to survey beacons and 'boundary makers' to the fences stopping people from freely walking the area (Brooks 2008, pp. 295–296). As reflected in the different meanings among Zulus, the nature reserve is a place that also evokes negative connotations.

Some of the communities on the northern border of the park were only moved there in the early 1960s. Evictions took place from nearby Empangeni to make way for an agricultural college (Save our iMfolozi Wilderness n.d.). For some current residents in African communities the park is therefore the outcome of multidirectional evictions.

In 1957, pioneering conservationist Ian Player founded the Wilderness Leadership School. Player, a former gold miner is credited with saving the white rhinoceros from extinction. The Leadership School provides wilderness trails through the park. The school prides itself on foreseeing 'the urgent need for a large body of well-informed, conservation-oriented leaders, capable of and dedicated to the defence of our planet's irreplaceable natural resources' (Wilderness Foundation Africa 2016). Brooks (2008, p. 301) reports how Player 'almost single-handedly imported a "cult of wilderness" from the United States into South Africa'. Interestingly, the Wilderness Leadership School was supported by the petroleum company Engen and the oil company Mobil. The trail experience started off being almost exclusively for whites – both South Africans and foreign tourists. This has changed to some extent today. There is no racist apartheid policy banning blacks from entering conservation areas, and some more diverse groups visit the park today. However, the park remains a place for middle-class and upper-class visitors. The large majority of interviewees from adjacent exclusively black communities who participated in this study reported that they had never set foot in the park. Some of the tourist facilities in the park, however, are co-managed by traditional chiefs from bordering neighbourhoods.

On the occasion of the centenary of the establishment of the park in 1995, Nelson Mandela, who became the first democratically elected President of South Africa, credited King Shaka with laying the foundation for the conservation efforts that led to the establishment of the Hluhluwe–iMfolozi Park. However, most of his speech looked to the future, addressing global competition in the tourism sector. Past evictions were not mentioned in his speech. Mandela showed his concern for the 'local people' who should be involved with the park so that they can also see the park as 'their heritage' (Mandela 1995). One can interpret Mandela's speech as an attempt to reconcile the history of conservation park with the existence of local residents by attempting to flesh out the economic opportunities that the park might bring to residents.

There are three dominant uses of the land today. These are as follows: high-end conservation and tourism; coal mining; peasant farming and living. I describe the specificities of the respective land uses and their compatibility and contradictions in turn. The fenced area of the Hluhluwe–iMfolozi Park houses all 'big five' animals (lion, leopard, rhinoceros, elephant, and buffalo), making the park particularly attractive for tourists. Game lodges and camps in 'pristine' environments guarantee overwhelming views of the Umfolozi River. The stay in the park is quite pricy. Holiday-makers can book lodges located in the park, with the most luxurious 5-star accommodations averaging £550 per night. Holiday-makers can also enter the park via one of the three gates using their own vehicle. Day passes are about a third of what is charged at the Kruger National Park, equating to around £5 per day. Driving through the park is only an option for those who own a car or those who can afford a 'Safari Tour' for around £70 (Hluhluwe Game Reserve, n.d.).

The park is both a conservation and a for-profit ecotourism enterprise that works under considerable budgetary constraints. Community liaison becomes an extra burden on a budget that is already under severe constraints. The relationship between park authorities and local community members remains strained. Local residents complain about the lack of job opportunities for locals and no contact person in cases of emergency when animals escape from the park (interview with activist, 27 March 2016). Employees working for the biodiversity conservation branch of the Department of Environmental Affairs (DEA) state that they lack the resources to liaise meaningfully with local communities. The park management in turn suffers from severe budget cuts. There was a 20% reduction in the budget during the 2016/7 financial year as well as a further reduction of almost the same margin for the following year (Breed 2016). Recent game auctions have also failed to fix budgetary holes (Carnie 2017). According to Ezemvelo staff members, this stressed budgetary situation has consequences for the capacity of the park management to engage the communities around the park (strategy meeting, 3 March 2017).

The second land use is traditional Zulu kettle farming and growing of maize, *amadumbe* (taro), and banana as well as rural resident area. All communities living around the park are headed by traditional chiefs ultimately reporting to the Zulu king. Household incomes in these areas are generally very low, but a large majority of people are surviving outside the cash economy (see Appendix). The majority of the households are female-led. Politically, the area is still very much aligned to the African National Congress (ANC). In the last general election, the ANC scored a comfortable majority of just over a 50% vote share in the iMfolozi Municipality, which, however, was below the provincial average of 55% (Electoral Commission of South Africa [IEC] 2019a, 2019b).

A third land use practice that takes place in the vicinity of Hluhluwe–iMfolozi Park is coal mining. Both existing coal mines are notorious for social unrest. Zululand Anthracite Colliery (ZAC) to the north of the park started anthracite (high-grade coal) production in 1987. It underwent frequent changes in ownership, with BHP Billiton and Rio Tinto being prior owners. Currently, the mine is operated by Menar, and its holding company is based in Luxembourg. Grievances of the local community centre around the lack of local employment opportunities, shortage of drinking water for the community, and dust pollution due to trucking (focus group discussion, 5 February 2016). ZAC was the site of frequent organised labour disputes. Go-slows, underground sit-ins, and destruction of property made headlines with The Guardian calling ZAC the 'ugly face of mining in South Africa' (Smith and Carrington 2015). According to an employee at the regional DMR in Durban:

> There are a lot of things that have happened since the mine started operating up until now: removal of people, water issues, people are complaining about the railway that was constructed, and it is not fenced although it is posing

threats to their livestocks and to the human beings as well. So, there has been quite a lot of protest.

(interview with staff member of Department of Mineral Resources, 13 April 2016)

The other mine at the other side of the park is Somkhele Coal Mine operated by Tendele Coal Mining, a subsidiary of the South African mining company Petmin. The company undertakes mining operations in Canada and South Africa. Somkhele Coal Mine started production in 2007, only five kilometres away from the Hluhluwe–iMfolozi Park's fence line.

Because of the clashes at ZAC and Somkhele over the years, mining earned a bad reputation in the area. Similar to the grievances voiced about ZAC, local communities around Somkhele complained about the lack of local employment opportunities and water shortages due to mining. Mining operations at Somkhele are also held responsible for frequent instances of lung disease and the cracking of houses (focus group discussion, 27 March 2016). Both developments had gone unchallenged by government conservationists at Ezemvelo. This changed when scoping applications for coal mining right next to Hluhluwe–iMfolozi Park surfaced in 2014. The Fuleni Mine proposal to mine only 40–100 metres away from the fence would displace up to six villages. For the biodiversity section of Ezemvelo, the mine proposal immediately raised a 'red flag' (focus group discussion, 10 February 2016). The reason for this was that the proposed mine would impact a high biodiversity priority area. An Ezemvelo Conservation Planner described the situation as follows:

> From Ezemvelo's perspective, we can't say no to every coal mine. We are commenting here and say: we don't see this as a compatible land activity. The mine is in the buffer zone of a protected area. . . . So what I am saying is that if the mine was in Dundee we wouldn't necessarily be involved. We are working with the community now to stop this mine because we have a common interest.
>
> (focus group discussion, 10 February 2016)

Historically, as we have seen in Chapter 3, conservation and mining were not generally seen as incompatible land-use practices. The alliance of Ian Player with Engen and Mobil shows that mining companies and conservationists can collaborate in principle. Both have something to gain. The fossil-fuel industry needs to improve their image with regards to their environmental footprint, and nature conservation is chronically cash-poor. Indeed, the former CEO of Ezemvelo highlighted, in an MoU with a subsidiary of Rio Tinto, the following:

> Whatever our varying interests, mining is a critical economic driver in South Africa. It makes absolute sense that conservation and mining work together to ensure development is pursued sustainably and in the combined interests of conservation, mining and community upliftment.
>
> (Ezemvelo KZN Wildlife 2014)

As long as mining is not taking place too close to conservation areas, these two land uses are compatible in principle. However, conservationists have also argued that the Somkhele Coal Mine already causes a nuisance. Only five kilometres away from the accommodation, noises and lights from the mine were disturbing holiday-makers at night.

A Moral Shock: Proposed Mining in Fuleni

The planned Fuleni Mine proposal envisages a mix of opencast and underground mining for high-grade coal. The project is proposed to last 32 years. The mining project is expected to initially create around 200 permanent jobs and an additional 400 jobs after about five years (Ibutho Coal 2015). Additionally, Ibutho Coal committed to spending around 6.6 million Rand (around £300.000) on basic educational projects, core business training, and other activities. A 5% share of the revenues was offered to a community trust (Carnie 2015).

The mine would have severe consequences for residents living in the project's footprint area. Ocilwane village lies in the centre of the project, and would disappear entirely. The village has about 230 households as well as a clinic, two schools, grazing land, three community graveyards, a worship site, and a water storage dam (Carnie 2015). Average households in Mfolozi comprise around five people. The majority of people living in Mfolozi are exceptionally young, with 68% of the local population being under 30 (Ibutho Coal 2014, p. 31). The average household cash income is very low in comparison to the national average household income. A number of interviewees for this study, however, pointed to the value of cattle and subsistence farming practices, highlighting that rural sustenance in potential mining areas would mostly not be derived from waged labour.

A crucial difference between the project and the existing coal mines is its location right at the border of the park. The Fuleni Mine would just start 40 to 100 metres from the fence of the conservation area. The geographic location of the proposed Fuleni Mine shapes political dynamics, as the alarm bells of conservationists started ringing once they heard of the project, thereby opening opportunities for alliance-building between the local community and conservation actors (see Figure 5.2). While the existing Somkhele Mine is situated a few kilometres from the wilderness area, the Fuleni Mine would just border the park.

In the application document, Ibutho Coal admits that there are potential impacts of mining taking place next to the Hluhluwe–iMfolozi Game Reserve, 'including visual and noise pollution, and [it] could affect its aesthetic and economic heritage significance negatively for the duration of construction and operation of the mine' (Ibutho Coal 2014, p. 27). Additionally, it is admitted that there would be adverse impacts on the quality of water of the Umfolozi River, even beyond the life of the mine. Archaeological gravesites would also be negatively impacted, as human remains would have to be removed or would be destroyed by mining activities. Especially from a spiritual point of view, this does not sit easily with the belief system of residents around the park, who have special relations with their ancestors. Places associated with oral tradition and landscape features relevant to

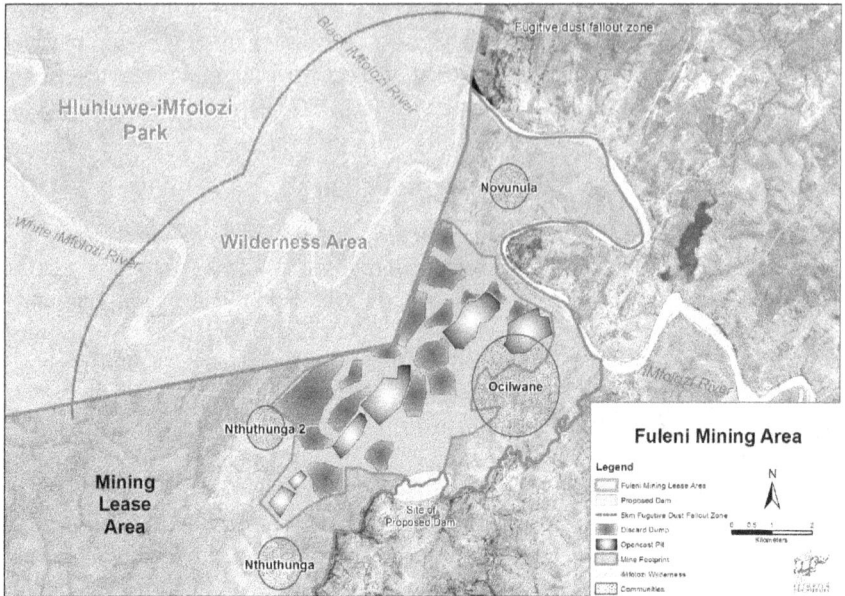

Figure 5.2 Map of proposed mining projects bordering the Hluhluwe–iMfolozi Park.

Source: Biodiversity Spatial Planning & Information, Ezemvelo KZN Wildlife, Pietermaritzburg, South Africa

local customs would be destroyed by mining activities. Ibutho, however, claimed that it would mitigate losses of traditional sites. However, possible impacts of mining were not sufficiently publicised with few people knowing about the mining proposal.

Social movement studies teach us that injustices alone do not cause resistance. Some people have to stand up and name injustices to bring awareness to a cause. Framing processes proposed by social movement studies help us understand the meaning-making processes in fossil-free social movements. Successful frames go beyond making people aware of threats and show ways to engage towards social change. Other than in the cases of the ZAC mine and the Somkhele mine, the Fuleni mine proposal was immediately resisted.

The Save Our iMfolozi Wilderness Leadership Group

The opposition to Fuleni mine is primarily organised by a leadership group of six activists forming the 'subcommittee opposing mining expansion' created to lead the Save our iMfolozi Wilderness campaign. The leadership group is a hand-picked group of people selected *ad hoc* by long-term conservationist and activist Sheila Berry who started the campaign by activating her activist network. Sheila's activism pioneered the field of wilderness psychology in the 1980s and introduced disadvantaged youth to the wilderness. In her activist work, Sheila found herself frequently clashing with the interests of the extractive industry. In her chosen field, she is a well-respected voice and activist.

The leadership group is a rather homogenous group of people. Of the six board members, five are white and have work experience in conservation-related non-profit organisations. The campaign's legal expert Kirsten Youens specialises in environmental litigation and has the experience of working on different development projects. Sifiso Dladla is the only black member of the leadership group. He joined the campaign a few months after the start of the campaign. His leadership had a profound influence on the frame-shift from a wilderness campaign to a focus on frontline communities, as we will see.

Much like most grassroots campaigns, SAVE is chronically underfunded. The subcommittee works on the campaign mostly on a voluntary basis. All the incoming funds are used for the campaign. Despite the constant worry around finance and progress of the coal mining project, the group must be described as confident and persistent. The website reports the following proudly: 'This campaign is seen to be one of the most bitterly contested battles in KwaZulu Natal's environmental history. It is also one of the most successful alliances between conservation and environmental and social justice' (iMfolozi Community Wilderness Alliance n.d.).

In the following section, I build on diverse sources that I collected for my study, including interview material with social movement activists, participant observations at activist meetings, protest events, focus groups, and journalistic sources. The frames presented in the following did not neatly succeed one another but overlapped to some extent. The periodisation in three phases is justified because of clear shifts of emphasis in the process of mobilisation.

As seen in Chapter 4, frames are mostly decided by movement leaders. In the following section, I discuss the master frames that the Save our iMfolozi campaign uses to oppose mining in Fuleni. As we will see, the leadership group first used the conservation area as a symbolic resource to make claims, thereby following a conservative tradition. The framing, however, shifted with the expansion of the leadership group to include a Sifiso who shifted the activist framing to demands of the frontline communities. The frame extension was finally precipitated by the moral shock of the assassination of an anti-mining activist on the Wild Coast in the Eastern Cape. This provoked the environmental justice (EJ) framing that links local struggles to global struggles against mining and other polluting practices.

Frame Setting: Starting a Conservation Campaign

[W]e didn't initially focus on the communities' rights in the beginning . . . it was very wilderness focused . . . so the whole impetus to it was to Save the Wilderness of iMfolozi from the mine . . . in fact, if you look at the website and you sort of look at the posts . . . go through [them, and] you will see these trends . . . it's all about wilderness, rhinos.

(interview with activist, 11 January 2017)

The SAVE campaign started in response to a moral shock: coal mining in Fuleni would destroy the iMfolozi Wilderness and harm wildlife in the oldest conservation park in Africa. The pristine wilderness area to the northern border of the proposed coal mining project is particularly dear to conservationists in South Africa but also globally. To frame the campaign as a conservation issue seemed like a natural reaction to the leaders and, most importantly, to Sheila Berry who came up with the campaign's name. The SAVE campaign immediately offered the convincing diagnosis that coal mining in Fuleni would irrevocably destroy the wilderness area. Coal mining was framed as a danger to the wilderness and the animals inhabiting it. The white rhino was particularly important to the campaign framing. The park management prides itself as the central conservation area responsible for saving the white rhino from extinction. The campaign urged the government to take a stance against the mining application in Fuleni. Conservation efforts are framed in the context of professionalism, care, and respect for nature. As the oldest preserved area of its kind in Africa, the park is an exemplar of skilful conservation management.

> The Park is a world leader in conservation research and practice. Its considerable number of innovations and contributions have significantly shaped the development of a highly professional and ethical system of management of natural wild areas that has been universally adopted by conservation agencies worldwide. Currently the Park supports the 'big five' and many other species of wildlife. It is the home of the founder population of the world's Southern White rhino, iconic animals. . . . All this could be changed and destroyed forever if the proposed Fuleni Coal Mine is given the go-ahead by the South African Government.
>
> (Save our iMfolozi Wilderness 2016)

Consequently, the SAVE campaign called on its 'natural' allies to oppose coal mining in Fuleni. An online petition was launched on the campaign website Avvaz.com 'to save the rhinos' on the very first day of the campaign, and the response illustrated the purchase of the conservation frame. At first glance, the response to the mining proposal was quite overwhelming. The text of the petition was supported by ten conservation-oriented groups in South Africa as well as the Gaia Foundation in London and WILD USA, a network that started organising conservation conferences back in the 1970s. The petition was soon signed by 50.000 people who shared the concern that the last sanctuary of the endangered white rhino could disappear: 'rangers fear it will destroy the wilderness and clear the way for poachers, who have butchered nearly 400 rhinos in South Africa this year alone but we can stop it if we act now' (Avaaz.org 2016). The idea of the petition was to raise the stakes and scandalise the mining proposal on the border of the conservation park: 'Let's make this mine publicly toxic. Click to join the urgent call to save the rhinos, and then tell everyone' (ibid.). The motivation of this framing was to spread the word and encourage target audiences to share relevant information.

The conservation frame also resonated with the international and local press. Only two months after the campaign started, the British Guardian ran an article echoing the sentiments of the campaign. Poaching of the white rhino and pollution from mining were presented as main threats to the conservation area (Smith and Barbee 2014). Moreover, the local press was favourably looking at the conservation campaign. They were mainly focusing on the plight of the wilderness, but also sharing the concerns of the communities surrounding the park.

Initially, the frame was only reluctantly supported by the state biodiversity conservation agency, Ezemvelo, which is managing the biodiversity of the park. As official conservationists mandated to protect biodiversity in the province, Ezemvelo conservation managers shared the diagnosis that coal mining would destroy the iMolozi wilderness area. This raised a 'red flag':

> From Ezemvelo's perspective, we can't say no to every coal mine. We are commenting here and say: we don't see this as a compatible land activity. The mine is in the buffer zone of a protected area. . . . So, what I am saying is that if the mine was in Dundee, we wouldn't necessarily be involved. We are working with the community now to stop this mine because we have a common interest.
>
> (interview with state conservationist, 10 February 2016)

The measured tone of this conservation manager was also reflected in the limited opposition they first showed against mining. They made little effort of their own to scandalise the mining proposal in the way the SAVE campaign did. The manager reported how her organisation was 'commenting' on and not 'fighting' coal mining. The conservation manager also suggested that mining opposition was not a policy of Ezemvelo, but that the mine proposal was procedurally not in line with zoning regulations. Their primary intervention was to submit comments through the official public participation channels. More confrontational repertoires of contention such as demonstrations were not taken into consideration (see Table 5.1).

Sheila Berry was frustrated with this attitude. In her activist life, she had seen how mining projects were proposed in close vicinity of conservation areas. In the campaigns she supported, there was always the feeling that resources in the subsoil had a higher value than the protected nature that covered it. In previous campaigns against sand mining in St. Lucia and against coal mining next to the Vaal River, Sheila had learnt how important a broad and sustained campaign is. Raising the stakes by raising awareness about the campaign among an interested public called for political leaders to take sides. The final word to end the St. Lucia mining proposal had former president Nelson Mandela himself vetoing the mine in St. Lucia. The experience from previous movement campaigns proved that mining proposals require a long and hard battle until they eventually get shelved.

In the case of Ezemvelo and the wider conservation community, Sheila Berry reflected that 'they are not warriors, they're advocates of the wilderness but they're not activists' (interview with activist, 19 January 2017). From her own experience, a mining campaign has to show the readiness to display acts of civil disobedience.

Table 5.1 Framing process of the iMfolozi Wilderness Campaign.

	Frame 1	*Frame 2*	*Frame 3*
Master frame	Save the wilderness/Save the white rhino	Listen to frontline communities: 'nothing about our land and us without us'	Environmental and social justice, anti-extractivism, fossil fuels, and climate change
Timing of the frame	First campaign weeks	Mid-2014 to early 2016	2016–today
Primary frame maker	Conservation community	Community-focused activists	Academics + conservation community + community-focused activists
Frame resonance	Conservation community + local and international press	Affected communities + environmental NGOs	National press + radical environmentalists
Limits to frame resonance	Exclusivity of the frame: White middle-class issue; limited grassroots mobilisation	Communities divided over coal mining; volatile political situation in the community	Fuzzy framing, leadership group losing control over framing (flexibility)
Corporate counterframing	Offsetting the wilderness; sustainable and modern mining technology	Creating jobs and development for the community	Limited public frame making

Source: Author's depiction

Despite its success to launch the campaign, the conservation focus limited the campaign, as it spoke to an audience that was either remote or not willing to engage on the ground. This limitation soon led to a frame shift to focus on frontline communities.

Limits of the Conservation Frame and Frame-Shifting

Despite the impressive resonance among members of the online conservation community and the press, crucial limits were seen as well. The first attempts to mobilise the conservation community pointed to a number of limits, especially regarding the racial composition of the target audience, the limited readiness of the target audience to engage in more confrontational political repertoires and the lack of motivational framing by the SAVE campaign.

The conservation community that was directly addressed by the SAVE campaign is not representative of average South Africans. As stated in the previous chapter, the conservation community in South Africa (and elsewhere) is largely middle- and upper-class white. Historically, the conservation community has been conservative and has also been built on exclusionary logics that discriminated against the black community (Martinez-Alier 2009). Even though there are some efforts to offer poorer black community members free walking tours in the conservation park bordering the proposed coal mine, conservation has very limited power to mobilise black communities. After apartheid, conservation remains a luxury that only a minority in society are able to enjoy as tourists. Rob Symons, a white farmer who runs the blog for the campaign, explained the strategic limitations of conservation campaigns.

> [T]here was a realization of course that wilderness: that's what they expecting us to defend and they basically would have kind of ignored it [the SAVE campaign]. Because what is this? A bunch of white guys, white people getting upset about a game reserve. So to think you know they could be easy dismissed as being white whereas mining is pushed as benefiting communities ... and communities have more weight than the wilderness area.
>
> (interview with activist, 11 January 2017)

As seen, the conservation frame reached the global conservation audience. Yet, in terms of grassroots political action, this had very limited local effects on mobilisation. In fact, the repertoire of contention was limited to signing an online petition and sharing content via emails and social media. For most people who sign online petitions, their engagement ends there. This was also the case for the 'Save the rhino' petition.

Another limitation was the structure of the frame itself. The framing of the petition lacked a call to action. Required action was limited to signing a petition and 'tell[ing] everyone'. Frames to save the rhino and the wilderness were linked to the faint hope that people would go beyond signing the petition. What exactly they were expected to do was not specified by the leadership group. It showed that place-focused activism, such as the defence of a specific nature reserve, needs local representatives and public appearances.

To sum up, the conservation frame was successful in raising awareness for the cause. When it came to grassroots mobilisation, the frame had limited effect. Having strong conservation arguments was not enough. With the wilderness advocacy groups taking a reluctant stance towards an activist campaign, Sheila Berry and the rest of the leadership group were forced to look elsewhere for support. This was counterintuitive, as 'initially, it was a wilderness issue' as Sheila Berry reports. The scope of the conservation frame was limiting, as it would address a largely middle-class, white audience. Conservation activists were relatively uncritical when it came to challenging the racial injustices under apartheid.

The frame shift from a conservation-oriented campaign to a social struggle for the rights of local communities can be attributed to the inclusion of the first black activist in the 'subcommittee for opposing mining expansion'. This shift offered insights into the deep fissures that the mining application had already opened in the community. The divisions that coal mining creates among communities became part of the campaign framing. Ultimately, the call for the inclusion of community demands amounted to a call for grassroots democracy: 'nothing about our land and us without us'. How did that shift happen?

Frame Shifting: Listening to the Frontline Community

> [M]ining processes go through different stages: exploration, mining, it is not just mining immediately . . . you have to get the rights, you then have to explore, then you have to sell the rights from exploration to mining rights, then you mine. All those processes – if done openly and democratically – communities should have a say in those. You could stop bad projects very early on.
>
> (activist interview, April 2016)

The SAVE campaign shifted the campaign focus within weeks after it started. Crucially, the frame shift was the result of strategic considerations within the leadership group. The conservation frame 'to save the iMfolozi wilderness' was initially useful to bring the struggle against coal mining in Fuleni to global attention. However, the conservation frame was weak in terms of starting a grassroots campaign showing the limitations of the top-down framing of the leadership group. The leadership group was shifting the frame to re-orient the focus of the campaign and become more sensitive to grassroots demands. The grassroots frame focused on the demand of 'nothing about our land and us without us'. The aim was to build a bottom-up grassroots movement by recruiting new activists and drawing on their knowledge. This involved activist learning processes, as the leadership group lacked practical knowledge concerning the local communities impacted by mining. The frame shift meant that the leadership group lost relative control over the direction of the campaign, as the social realities in Fuleni were largely unknown to the group. At the core of the grassroots framing was the realisation that seven villages would be evicted and 16.500 villagers directly affected by the coal mine in Fuleni (Save our iMfolozi Wilderness n.d.).

The frame shift coincided with a change in leadership composition. The incorporation of Sifiso Dladla in the leadership group shifted the framing of the campaign. Sheila Berry remembers the moment Sifiso Dladla became active in the leadership group:

> Initially, it was a wilderness issue for me, but Sifiso, bless his heart, immediately had to go to Fuleni. He said, 'Sheila, I've gotta go, I'm gonna find out what's going on there'. So I said, 'Go, go, go', and he took my car, and off he goes. And goes to a meeting, attends meeting . . . [being, sic.] a very strong voice saying, 'No, we don't want the mine'.
>
> (interview with activist, 19 January 2017)

As a journalist-turned-activist, Sifiso Dladla had the communication skills and the ambition to become a mouthpiece for mine-affected communities. In his career, he took different positions in NGOs, as he was acquainted with the worlds of both professional NGO circles and grassroots activism. In his activist work, he focuses on the social impacts of mines on communities. This also made him the first member of the leadership group with first-hand insights into the political dynamics around mining. The orientation of the leadership group was subsequently shaped by the experiences of Sifiso Dladla in Fuleni and his prior knowledge of mine-affected communities. Commenting on the power dynamics around mining in rural South Africa, he sums up the following:

> Mining in South Africa always or mostly happens in rural communities where there is [sic.] traditional leaders; the public participation process is always flawed because the mining company consults the traditional authority and then it [traditional authority] takes the decision for the community.
> (interview with activist, 30 June 2015)

The role of traditional authorities is criticised along with that of elite politicians who sustain very close relations with the mining industry. Sifiso Dladla characterises the mining sector as a violent and latent threat to social peace in South Africa. While the conservation framing was focusing on unleashing violence upon the animal world and the destruction of pristine nature, a community-led framing emphasises violence against local people.

> Politically, it's very sad that the high-ranking politicians are slowly buying into mines. If you look at the President's family, they have [an] interest in mining, [and] if you look at the speaker of parliament: mining industry. The deputy president: mining industry. The top ranks of the politicians – they have [an] interest in mining. Therefore, the result was Marikana. If you look at it socially, mining communities are crying, people are being killed day in and day out. The change of social fabrics . . . there is no peace wherever there is mining.
> (interview with activist, 30 June 2015)

Sifiso Dladla's intellectual leadership in terms of the politics of mining in South Africa and the plight of mine-affected communities was a key in changing the political direction of the campaign. Notably, Sifiso Dladla remains the only one who is able to speak to key political actors in their native Zulu. Thus, he is the only member of the leadership group who is able to communicate with all community members at meetings. Sifiso Dladla established close links with the activist leaders living in the villages that the mining company planned to evacuate for mining operations.

One of the main challenges of framing is to diffuse frames to reach target audiences. This was particularly difficult in a rural setting such as Fuleni. Most community members there had no idea about the mining application and the possibility of their future eviction. Neither did community members know about the general effects of mining on community life. The very lack of information on the mine project became a part of the activist framing. Especially older community members who had experienced apartheid feel a frustration among rural communities that they are not appropriately involved in political decision-making and that their voices are frequently ignored.

The first public consultation meeting called by Ibutho Coal was an opportunity for the leadership group to learn about local dynamics in Fuleni, forge alliances with activist groups, and understand local grievances. Around 100 community members had gathered in Ocilwane (see Figure 5.2) to hear what the representatives of Ibutho Coal proposed to them. At the first public consultation meeting in Ocilwane on 31 August 2014, a local leader vocally opposed the mine, ultimately forcing company representatives to leave. Community activists showed their particular discontent with the way they were addressed by the mining company's representatives. Addressing Ibutho Coal's Director Thembi Myeni directly, community activist Phila Ndimande presented his opposition to the mining project in the following terms:

> Do not call another meeting. Ocilwane does not want the mine no matter what you offer us, so do not come back. Today you have heard the people complain that you are ignoring our voices and we are being undermined and disrespected. You know we speak Zulu, yet you come to the community with a presentation in English. You are not welcome to come again.
>
> (Mining Weekly 2014)

The meeting became tumultuous after the scope of the mining project was outlined: the mine would displace homes and change the community forever. The confrontation led the representatives of Ibutho Coal to call off the meeting that day after Phila Ndimande told them to leave. The premature end of the meeting sent a message to the rest of the community that mining was not inevitable. With both Sifiso Dladla and Sheila Berry being present at the meeting, the SAVE campaign was also able to credibly reach out to the community.

The public display of opposition shaped the frame to portray Ibutho Coal as lacking respect for and taking decisions without properly engaging the communities. As seen in the previous chapter, mining applications in South Africa require applicants to hold public consultation meetings with the impacted and affected communities. The occasion of the consultation meeting allowed community members to show their disapproval of the mining project and win a symbolic victory. The participation meeting was also instrumental in finding new grassroots activists who wanted to do some groundwork and raise awareness in the community.

The accusation that Ibutho disrespected the community was taken further, as Ibutho was perceived to cut corners to accelerate the mining project. Half a year after Ibutho had to call off the public participation meeting in Ocilwane, the company pasted what had been identified as demolition stickers on the homes of residents.

The subcommittee answered, by pointing to the psychological strains, the proposal brought to the community and the irreversible damages of mining operations for communities. Commenting on the stickers for the press, Sifiso Dladla raised questions concerning the moral integrity of the company.

> How can they have the audacity to put these stickers on people's homes without a proper consultation process? How will these people be able to sleep now that they know they have been identified for relocation? Where will they go to? What is the psychological effect of finding one of these stickers on your home? What compensation will they get for the loss of their homes, their grazing land, ploughing land, ancestral graves and other infrastructure in their communities?
>
> (Carnie 2015)

Members from the subcommittee agreed that Ibutho wanted to demoralise community members and make them feel that a decision in favour of mining had already been taken. The subcommittee's legal expert Kirsten Youens commented the following: 'What Ibutho is doing is acting as though its obtaining rights over Fuleni Reserve is a foregone conclusion and that any consultation with the community is not necessary' while also threatening legal action (Carnie 2015).

Limits of the Community Frame

Even though Ibutho's reputation in the community suffered, mobilisation against mining using the frontline frame faced several constraints. The rural political power structures around Fuleni created obstacles to local mobilisation. The role of the traditional authority working with the mining company especially limited the efforts to mobilise the grassroots. Behind the scenes, Ibutho Coal seemed to be working hard to get the traditional authority (*chief*) of the community on their side. The elderly chief of the community, who many community members still accept as the head of the community, avoided a discussion on mining with activists. Instead, the chief was increasingly seen to be in collusion with Ibutho Coal. The chief seemed less and less available to discuss matters with community members, causing a sense of alienation:

> this [mining] consultant guy driving his car [to] our chief's house. It is making us useless because that guy is like a first-born in our chief's house. He goes wherever he wants to go. But if I go there, I become a stranger and he asks me what I want.
>
> (interview with activist, 27 March 2016)

The chief's stance on mining imposed limits on the grassroots framing of protest, as the mining proposal was partly overshadowed by adherents and opponents of the traditional leadership. Community members feel that they expose themselves if they go against the will of the chief. An activist reported how they were told by her chief to stop their activism and subsequently felt threatened by the consequences of their activism.

Another limiting factor of the community-led frame was that the emergent figure of the anti-mining activist was contested and often misunderstood. A young activist reports that her fellow community members think that she receives payment for her activism because it brings her in contact with white activists: '[P]eople think that being an activist. . . . I get paid . . . because of the skin colour, because you guys are white and I am black' (interview with activist, 27 March 2016). Both the strained relationship with traditional authorities and racial prejudices within her community make it hard for her to engage in activism.

The Save/SAVE leadership group allocated resources for community members to learn about mining and strengthen bottom-up resistance to the mining project. Exchange visits bringing community members in Fuleni and other villages to other mine-affected like Marikana or the Witbank were organised. Local and national NGOs collaborated with the SAVE leadership group to plan these trips. Exchange visits helped to view the mining proposal in Fuleni from a broader perspective and strengthen the arguments to oppose mining. Impressions from mine-affected communities elsewhere in South Africa helped activists to take a stance.

> In Mpumalanga, when you reach the age of 20, you already got asthma or TB due to the air pollution. So you can't work in the mining industry when you already got asthma or TB' cause you've already been sick. But the cause is the same mine. So that is when we said that we don't want to hear nothing about that mining because it is going to be killing us. It is not gonna help us with anything. And when they say they are coming with job opportunities . . . it is not for all of us in the villages. It is for those who are going to benefit and the rest must suffer. So we said no to all of that.
>
> (interview with activist, 27 March 2016)

Grassroots learning and forging of new activist connections to NGOs and activist-scholar networks amalgamated into the extension of the community frame. Widening the frame of resistance added new perspectives for such mining communities more generally. With other actors getting involved against the mining proposal in Fuleni, demands to end the extraction of fossil fuels and stop climate change were embraced.

Frame Extension: Anti-Mining, Fossil Fuels, and Climate Change

Tragically, the frame extension was a reaction to the assassination of anti-mining activist Sikhosiphi 'Bazooka' Rhadebe on the Wild Coast. In March 2016, Rhadebe was assassinated in front of his home. He had been at the forefront of a movement against proposed titanium mining. The Australian MRC minerals company applying for the right to mine the Wild Coast subsequently denied any involvement in the murder of Rhadebe (Rogers 2016). The circumstances of the assassination remain unresolved. The moral shock of the assassination led to the consolidation of environmental movements standing up against mining in South Africa. The funeral was attended by activists, church leaders, political parties, and SAVE campaigners. The shared grievance and anger around the loss of the activist lead to the collaboration of activist groups and strategic discussions on how to link the demands of different social movement actors addressing the plight of rural South Africa.

The SAVE campaign leadership worked closely with socio-environmental NGOs and scholar-activists from the start of the campaign in 2014, but their intellectual input to the campaign framing was only included more decisively in early 2016. Demands for social and environmental justice were circulating in NGO circles and among scholar-activists in institutions of higher education for some time. But the campaign framing was only extended when external shocks led the leadership group to articulate broader demands to take climate change action and end excessive fossil-fuel extraction. Embracing broader demands beyond the Fuleni resistance against coal mining also meant that the leadership group allowed more actors into the circle of frame making. Henceforth, the SAVE campaign became more deeply inserted in the loose socio-environmental activist network in South Africa.

After the death of Rhadebe, for a moment, the political opportunity structure looked like it could change so that the political system would have to make concessions for the social movements' claim-making. The violent act drew media attention and social movement's claim-making received more attention. The assassination was a key event for activists, as the issue of mining was discussed on television, in newspapers and in the parliament.

A picket was staged at the KZN provincial Department of Mineral Resources (DMR) around one month after the funeral in Durban. Subsequently, a public debate and a press conference were held at the Centre for Civil Society, which provided a platform for the leading voices of anti-mining activists. Patrick Bond, the former director of the Centre for Civil Society at the University of KwaZulu-Natal, also became more involved in shaping the frame of the campaign. The press release that was issued that day pointed beyond planned coal mining in Fuleni using resistance to fossil fuels as a frame to connect local struggles to failures in the political system at large:

> In addition to the curse of coal mining, KZN is further threatened by fracking and off-shore drilling for oil – fossil-fuel extraction that the government committed our country to decreasing by signing international protocols to reduce greenhouse emissions.
>
> (Jolly 2016)

The commentary on national television that day by activists from around the country pointed to the unsustainability of mining and the call for the government to review all mining licences involving fossil fuels. The panel included speakers from different mine-affected communities and communities that might host mines in the future. The press statements by the SAVE leadership team released on the occasion also reflected the attempt to link the local struggle to global environmental processes. In a press statement quoted in a Sunday newspaper, Sheila Berry reflected on the hope to broaden the alliances against exploitative social practices more generally.

> Given the enormous challenges facing the planet, there is an imperative for like-minded people and civil society and NGOs to work together to achieve our goals of a better quality of life and to wage war against greed and exploitation, particularly when it affects the environment and those with few resources to fight back.
>
> (Clarke 2018)

The call for a broad civil society alliance against environmental and social exploitation was a more explicit call for cross-societal coalitions than earlier frames. 'Waging war' was more belligerent than earlier calls to 'click to join the urgent call to save the rhinos, and then tell everyone', which was embraced when the conservation frame was in use.

The environmental and social justice frame bridges the two previous frames. It situates the campaign between the demands for social and environmental action. The language that environmental and social justice framing embraces is more radical than the previous framing. Environmental and social justice questions the capitalist system by proposing a politics that tackles racial and economic inequalities. In the conclusion of this chapter, I reflect on how the campaign's framing started with a rather conservative frame to later include more grassroots demands and finally broadened its scope.

Conclusion and Discussion: Frame Processes in the iMfolozi Wilderness Campaign

Compared to other grassroots environmentalist groups that dissolved after a shorter period (Staggenborg 2020, pp. 33–63), the SAVE campaign managed to keep the campaign alive. The campaign stayed in touch with social reality, as it showed flexibility in shifting the campaign's framing when the conservation framing showed its limits. Diversification of the leadership team allowed the campaign to embrace new positions more in line with grassroots experiences. Experimenting with different action frames also helped to activate organisations that previously had little interaction with conservation activists.

The SAVE campaign attained global attention with the help of the online petition to 'save the rhino', which helped to bring the struggle against coal mining to a worldwide conservation audience. Conversation groups from the US and Europe endorsed the struggle. However, the majority of supporters remained passive bystanders and observed from a distance. The initial call to 'save the rhinos' lacked the motivational framing to mobilise people beyond online endorsements and information sharing. Despite its resonance with the international conservation community and international media, the conservation frame was largely associated with white middle-class concerns and focused too narrowly on online activist repertoires. Frustrations among members of the leadership group were channelled into changing the direction of the campaign. While the conservation frame was crafted by its founder Sheila Berry as a kneejerk reaction to the moral outrage of the coal application, the community frame was developed in conversation with residents from outside the park.

The leadership group showed flexibility by shifting away from its initial conservation framing after just a few weeks of campaign action. Key to the frame shift from a conservation to community-led framing was a change in the composition of the leadership group – the inclusion of the first black leader Sifiso Dladla. Diversifying the leadership group led to a significant change in political direction.

Bottom-up framing shifted the emphasis of the campaign to endorse more grassroots concerns, which increased the mobilisation potential and changed the repertoires of contention to be more confrontational. The emphasis of the campaign framing moved outside the conservation park to focus on the surrounding communities where a number of protests were supported. Protest marches and disruptions at public participation meetings were supported and co-organised. Symbolic victories against the mining company Ibutho Coal were won when meetings were called off.

The frame shift resonated with community members, as it took their grievances seriously. Particularly, the apparent disregard for local customs and livelihoods and the flawed public engagement by Ibutho Coal found expression in the frames used by the SAVE campaign. Action frames are particularly successful when they sufficiently engage with the 'existing cultural repertoire' (Hewitt and McCammon 2005, p. 34).

However, there were clear limits to the use of the community frame too. As an outside actor, the campaign team learnt about divided communities and the intricacies of rural politics led by traditional authorities. Many activists were constrained from participating in activism. They were intimidated in order to end their resistance. The campaign was also unable to mobilise large crowds because of the geographic dispersion of residents and limited campaign funding. With Sifiso Dladla being the only campaigner with prior knowledge of the community, the campaign only had limited and partial success in the communities.

The orientation towards the environmental and social justice frame was triggered by the assassination of an anti-activist on the Wild Coast. The assassination opened a window of opportunity to broaden the campaign framing to embrace wider demands, such as fossil-fuel resistance and climate change. This coincided with the leadership group taking a backseat in influencing the framing of the campaign. NGO professionals and scholar-activists were influencing drafts of memoranda and newspaper opinion pieces that were reproduced on the campaign website.

The demand to stop fossil fuels and address climate change was only embraced late in the campaign. In other words, activists embraced the planetary implications of their actions after they had sufficiently explored the possible local implications of mining. The scale of demands and ambition moved upwards. Framing the campaign more decisively around fossil resistance and climate change opened up the possibility to speak the same language as other movements nationally and globally. Strictly speaking, only the third frame explicitly linked the mining application to fossil-fuel opposition. While the first two frames were relating to particular local struggles, environmental and social justice framing elevated the coal resistance to be more in line with other social struggles against mining. EJ originates from the US context where it was first mobilised to fight polluting industries such as incinerators in predominantly Latino and black communities. EJ later also became a more analytical frame to study the relation between disadvantaged communities and high levels of pollution (Martinez-Alier 2009). I will come back to the EJ frame in Chapter 7 to the potentials for global fossil-free mobilisation.

It appears that EJ becomes more popular after a longer period of obscurity. The concept of EJ was first used in the South African context at the beginning of the 1990s. In his book on *Environmental Justice in South Africa*, McDonald (2002, p. 1) states that 'flora and fauna were often considered more important than the majority of the country's population'. Since 2012, the Mining and Environmental Justice Network of South Africa (MEJCON-SA) has started to promote the environmental and human rights of mine-affected communities, train community members, and support community activists. MEJCON-SA works with more than 300 groups throughout the country, including the SAVE campaign.

By embracing the EJ frame, the SAVE campaign opened up to an ecology of organisations that is presented with similar political dynamics. The activist network helps to access information, expertise, and funding opportunities. However, because of the numerous environmental struggles throughout the country, the everyday organisation of resistance rests on the shoulders of the local leadership team. As shown, SAVE activists managed to extend their network and broaden the scale and scope of the campaign.

References

Avaaz.org, 2016. *Save the rhinos.* Available from: https://secure.avaaz.org/campaign/en/hinos_worst_neighbour_sa_7/

Breed, R., 2016, March 18. Big budget cuts for Ezemvelo. *Zululand Observer* [Empangeni]. Available from: https://zululandobserver.co.za/104684/big-budget-cuts-for-ezemvelo/

Brooks, S., 2005. Images of 'Wild Africa': nature tourism and the (re)creation of Hluhluwe game reserve, 1930–1945. *Journal of Historical Geography*, 31 (2), 220–240. doi:10.1016/j.jhg.2004.12.020

Brooks, S., 2008. Royal precedents and landscape midwives: Claiming the Zululand wilderness. In B. Carton, J. Laband and J. Sithole, eds., *Zulu identities: Being Zulu, past and present.* Pietermaritzburg, South Africa: University of KwaZulu-Natal Press, 293–303.

Carnie, T., 2015, March 18. Coal mine's effects would be huge. *Business Report.* Available from: www.iol.co.za/business-report/economy/coal-mines-effects-would-be-huge-1833756

Carnie, T., 2017, May 22. Wildlife auction brings in R8m for the struggling Ezemvelo KZN wildlife. *Business Live.* Available from: www.businesslive.co.za/bd/national/

science-and-environment/2017-05-22-wildlife-auction-brings-in-r8m-for-the-struggling-ezemvelo-kzn-wildlife/

Clarke, C., 2018, August 6. iMfolozi: war against mine greed. *The Sunday Tribune*. Available from: http://earthlorefoundation.org/imfolozi-war-against-mine-greed-the-sunday-tribune/

Climate Action Network Europe, 2021. *About us*. Available from: https://beyond-coal.eu/about-us/who-we-are/

Electoral Commission of South Africa, 2019a. *2019 National elections, KZN281 – uMfolozi*. Available from: www.elections.org.za/content/NPEPublicReports/699/Results%20Report/KN/KZN281/KZN281.pdf

Electoral Commission of South Africa, 2019b. *2019 national elections: all provinces*. Available from: www.elections.org.za/content/NPEPublicReports/699/Results%20Report/National.pdf

Environmental Justice Atlas, n.d. *Coal*. Available from: https://ejatlas.org/commodity/coal

Ezemvelo KZN Wildlife, 2014, December 9. *The coincidence is deeply ironic: at a time when people in South Africa and worldwide reflect on the loss of one of the world's most celebrated conservationists, Dr Ian Player, frantic efforts are being made to save* (Facebook status update). Available from: www.facebook.com/EzemveloKZNWildlife/posts/855476311139551

Gellert, P.K. and Ciccantell, P.S., 2020. Coal's persistence in the capitalist world-economy: against teleology in energy 'transition' narratives. *Sociology of Development*, 6 (2), 194–221. doi:10.1525/sod.2020.6.2.194

German Institute for Economic Research; Wuppertal Institute for Climate, Environmental and Energy, Ecologic Institute; Energy Economics, 2019. *Phasing out coal in the German energy sector*. Available from: https://epub.wupperinst.org/frontdoor/deliver/index/docId/7265/file/7265_Phasing_Out_Coal.pdf

Greenpeace, 2008. *The true cost of coal*. Amsterdam: Greenpeace International.

Guy, J., 1982. *The destruction of the Zulu Kingdom: the civil war in Zululand, 1879–1884*. Johannesburg, South Africa: Ravan Press.

Hewitt, L. and McCammon, H.J., 2005. Explaining suffrage mobilization: balance, neutralization, and range in collective action frames. *In*: J.A. Noakes and H. Johnston, eds., *Frames of protest: social movements and the framing perspective*. Lanham: Rowman & Littlefield, 33–52.

Hluhluwe Game Reserve, n.d. *Hluhluwe accommodation rates & bookings*. Available from: https://hluhluwegamereserve.com/hluhluwe-game-reserve-accommodation/

Ibutho Coal., 2014. *Fuleni anthracite project* (Draft Scoping Report). Available from: www.ibuthocoal.co.za (website no longer available)

Ibutho Coal, 2015. *Final scoping report*. Available from: www.ibuthocoal.co.za (website no longer available)

iMfolozi Community Wilderness Alliance, n.d. *About*. Available from: https://saveourwilderness.org/about/icwa/

Jolly, T., 2016, April 20. Solidarity among anti-mining groups. *Zululand Observer*. Available from: https://zululandobserver.co.za/109371/solidarity-among-anti-mining-groups/

Mamdani, M., 1996. *Citizen and subject: contemporary Africa and the legacy of late colonialism*. Princeton, NJ: Princeton University Press.

Mandela, N., 1995. *Address by President Nelson Mandela at the centennial celebrations of the Hluhluwe-Umfolozi Park and Greater St Lucia Wetland Park, Hluhluwe Game Reserve, KwaZulu-Natal*. South African Government website. Available from: www.mandela.gov.za/mandela_speeches/1995/950430_hluhluwe.htm

Martinez-Alier, J., 2009. Environmental justice in the United States and South Africa. *In*: M. Reynolds, C. Blackmore and M.J. Smith, eds., *The environmental responsibility reader*. London and New York: Zed Books, 247–255.

McDonald, D.A., 2002. *Environmental justice in South Africa*. Athens, OH: Ohio University Press.

Mining Weekly, 2014, September 4. Fuleni community send Ibutho coal packing. *Mining Weekly*. Available from: www.miningweekly.com/article/fuleni-community-send-ibutho-coal-packing-2014-09-04/rep_id:3650

Rogers, G., 2016, April 11. Company denies murder link. *Herald LIVE*. Available from: https://www.heraldlive.co.za/news/2016-04-11-company-denies-murder-link/

Save our iMfolozi Wilderness, 2016. *The wilderness and Hluhluwe-iMfolozi Park* [Blogpost]. Available from: https://saveourwilderness.org/about/the-wilderness-and-hluhluwe-imfolozi-park/

Save our iMfolozi Wilderness, n.d. *Fuleni community*. Available from: https://saveourwilderness.org/about/fuleni-community/

Smith, D. and Barbee, J., 2014, July 17. Mining poses new threat to world's greatest rhino sanctuary. *The Guardian*. Available from: www.theguardian.com/environment/2014/jul/17/-sp-mining-threat-south- africa-rhino-sanctuary-poaching

Smith, D. and Carrington, D., 2015, May 26. Dust, TB and HIV: the ugly face of mining in South Africa. *The Guardian*. Available from: www.theguardian.com/environment/2015/may/26/dust-tb-hiv-ugly-face-mining-south-africa

Staggenborg, S., 2020. *Grassroots environmentalism*. Cambridge: Cambridge University Press.

Wilderness Foundation Africa, 2016. *Wilderness leadership school*. Available from: www.wildernessfoundation.co.za/global-network/wilderness-leadership-school

6 Movement Tactics for a Frack Free South Africa

As seen in the previous chapter, SAVE has repeatedly amended its framing in order to mobilise activists and adapt its frames to new circumstances. Notably, changing frames has been a part of the learning processes of social movement leaders. Adaptiveness and the adoption of cultural scripts helped them to steer the campaign and remain relevant over time. The composition of the leadership group had a crucial influence on the framing of the protest. The movement also widened the potential audiences, since EJ is used around the world as we will see later in the global outlook.

This chapter deals with the ways in which movement tactics are formulated in fossil-free struggles. Counter-hegemonic fossil-free movement tactics aim to counter the claims made by the industry or the state and obstruct the application procedure physically. First, I describe the global context in which the new resource frontier around fracking is opening. I then look at Frack Free South Africa (FFSA) challenge to the oil and gas industry. Their observed tactics range from blocking and delaying, educating, and connecting to prefiguring that will describe in turn. As I briefly discuss in the following section, the existing body of literature is mostly concerned with claim-making against fracking in the context of the Global North. We lack an understanding of movement tactics in semi-peripheral countries that could potentially become the next fracking frontier.

As of May 2022, the EJA lists 48 global resistance movements to fracking on all continents except Australia (EJA n.d.). Resistance movements typically involve residents, local, national or international NGOs, farmers, and, in some cases, unions or political parties (ibid., n.d.). In-depth studies on fracking resistance, however, look overwhelmingly at cases from North America (e.g. Ladd 2018; Vasi *et al*. 2015; Fusco and Carter 2017; Temper 2019; Neville and Weinthal 2016) and Europe (e.g. Munci 2019; Milhaylov 2020; Nyberg *et al*. 2018; Vesalon and Creţan 2015) (see also Figure 6.1). This bias needs to be addressed, as resistance movements in other parts of the world oppose an industry interested in large estimated shale gas reserves in countries including Argentina (11% of the world's total shale gas), Algeria (9.7%), Mexico (7.5%), and South Africa (5.3%) (Energy Information Administration 2015). These countries are the next fracking frontiers (Hadad *et al*. 2021; Azubuike *et al*. 2018).

DOI: 10.4324/9781003110835-8

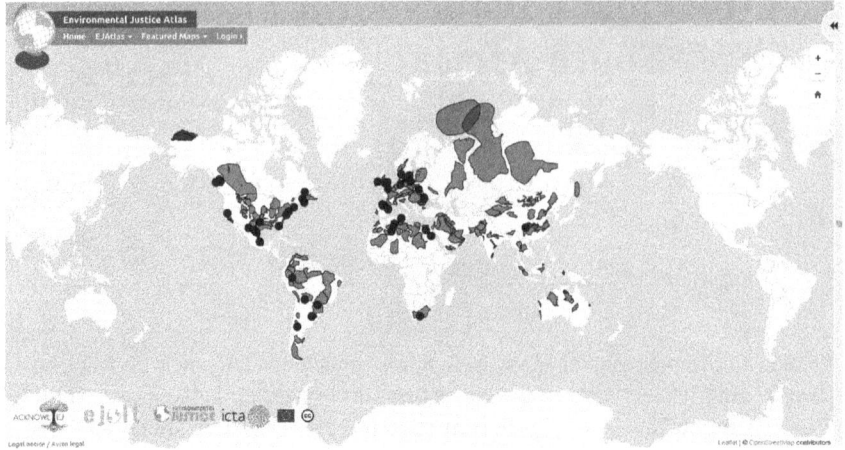

Figure 6.1 Global shale basins in grey with black dots highlighting cases of resistance to fracking.

Source: Environmental Justice Atlas 2022

The state is a central actor in the exploitation of natural resources (Collier 2010). The South African Deputy Minister for Mineral Resources claimed in parliament that fracking will be 'a true game-changer for our energy mix in South Africa' (FFSA 2016a). The emerging fracking frontier in developmental states like South Africa is often framed around the national interest. In South Africa, a central promise of the ANC platform is to roll out public services such as electricity and tap water supply to poor households (ANC 2019). After Mandela's tenure ended in 1999, the delayed roll-out of service delivery sparked widespread with poorer citizens asking for electricity, water and sanitation protests around the country (Wasserman *et al.* 2018; Alexander 2010; Bond 1998). In this context, fracking is proposed as a developmental fix to address the crises of energy shortages, mass unemployment, and even environmental pollution. ANC politicians and corporate leaders, therefore, link the potential of fracking gas to the prospects of energy autonomy, jobs and cleaner and cheap energy. Like the anti-coal movement discussed in the previous chapter, the frack-free movement in South Africa stands against a relatively stable fossil-fuel bloc.

However, with fossil-fuel hegemony dwindling globally in the face of visible climatic changes and with aggravated water crises in South Africa, fracking projects provoke resistance. Even though the ANC receives stable majorities in national elections, the party faces considerable headwind from extra-parliamentary groups that are discontent with rampant inequality in the country (Alexander 2010).

In 2011, the multinational corporations Shell and Falcon and some South African exploration companies applied for exploration rights in the semi-arid Karoo desert. Four years later, exploration applications, mostly by smaller oil and gas companies in the eastern provinces of KZN, Mpumalanga, and the Free State provoked the emergence of FFSA. Treasure the Karoo Action Group (TKAG) was previsouly established to fight fracking in the Karoo. The focus of this chapter is on the eastern applications and the activist tactics of FFSA. FFSA was formed when 3.2 million hectares of land were applied for shale gas exploration in 2015. As I show in this chapter, the prospect of fracking in the region sparked a moral shock. Subsequently, FFSA practised four different movement tactics to both consolidate the movement internally and to voice opposition in public.

Considering that anti-fracking movements are at the forefront of challenging an industry that would significantly alter current land-use practices in vast swaths of South Africa, the lack of scholarly attention seems odd (Finkeldey 2018). Previous studies discussed petroleum regulation's readiness for the gas industry (du Plessis 2015), integrated planning (Scholes *et al.* 2016), technical readiness (Academy of Science of South Africa 2016), and multi-level governance (Atkinson 2018) linked to fracking in South Africa; there has been little work on grassroots organisations fighting the industry. Public claim-making and the inner organisational life of social movements mobilising against fracking in South Africa have not been explored in any depth so far.

Anti-fracking movements in South Africa expose a number of risks of fracking, such as water pollution, climate change and boom-and-bust economies, thereby presenting a radically different story when compared to those propagated by industry and political elites. The fight over fracking is characterised by the uneven distribution of resources. The industry and ruling party have media attention and resources to commission studies to make their arguments widely public while social movements have to raise funds and create political opportunities to be heard in public. Yet with growing discontent globally and in South Africa about fossil fuels, the South African 'fossil-fuel bloc' has come under stress. In this context, FFSA managed to build a sustained opposition movement. This movement was formalised and managed to make a number of public appearances.

FFSA uses four activist tactics to stop oil and gas companies to drill for gas, namely blocking and delaying, educating, connecting, and prefiguring. Existing studies on fracking resistance discuss the movements' scaling-up tactics (Neville and Weinthal 2016), educational initiatives (Vasi *et al.* 2015), or blocking tactics (Temper 2019). So far, such tactics have been considered mainly in isolation from each other, and their analysis often serves to support an argument concerning only the specific tactic or aspect of a movement.

FFSA's struggle needs to be understood against the backdrop of the crisis of the global 'fossil-fuel bloc' and the struggle of the ANC to add fracking gas to the South African energy mix to fix developmental backlogs in South Africa. Looking at a social movement's counter-hegemonic tactics renders visible both the internal sense-making process as well as the tactics to face the opponent. In the case of FFSA, three out of four tactics aim at building the movement internally and gaining strength to increase numbers. The blocking and delaying tactic confronts the oil and gas industry in face-to-face interactions as well as in the written form. *Educating* the community is important, as fracking is not well known in South

Africa and the information provided by oil and gas companies is one-sided. *Connecting* with other civil society actors, the media and political parties have the potential to strengthen the movement and make it known to the wider public. By *prefiguring* the kinds of communal practices and livelihoods, FFSA is celebrating the existing land-use practices in order to counter arguments for land-use changes as a result of fracking.

This chapter proceeds in the following steps: first, the fracking applications in KZN and the neighbouring provinces were a moral shock to a number of environmentally minded people who became anti-fracking activists and founded FFSA. They also encountered particular challenges for social movement building linked to fracking applications. Second, I present the leadership group of FFSA and the stance they take against the industry. In this section, I also discuss the place history of the area where most activists come from. Third, I discuss movement tactics against fracking. Finally, the movement's political opportunities are discussed in light of the efforts of national elites to link a possible gas boom to the national interest.

A Moral Shock: Potential Fracking in KwaZulu-Natal

Exploration applications for unconventional gas deposits mushroomed in KZN and the neighbouring provinces around 2015. The central corporate actor in the eastern parts of South Africa is a Texas-based exploration company called Rhino Oil and Gas.[1] Rhino Oil and Gas was established in 2012 with the goal of acquiring oil and gas concessions in Africa. In Africa, the company is pursuing oil and gas concessions in Namibia, Senegal, Guinea-Bissau, and Comoros (Rhino Resources 2018). In South Africa, Rhino is currently proposing to exploit gas resources in a total of one offshore and five onshore applications.

In the northern parts of KZN, a number of corporations are competing for market shares for an anticipated gas boom. The scoping report for an Environmental Impact Assessment (EIA) for exploring the economic viability of gas development in the northern parts of KZN was submitted by Rhino in July 2016. The exploration application was accepted by the relevant authority, the Petroleum Agency of South Africa (PASA), one month later. The Australia-based company Kinetiko Energy lodged an application in 2016 for southern parts of the Mpumalanga province and the northern parts of KZN, stating that they are determined to develop 'clean energy solutions in a high gas demand region and create jobs' (Kinetiko 2013). A further application submitted by Sungu Sungu for the area was withdrawn in 2017 (African News Agency Reporters 2017). The Rhino Oil and Gas exploration area around Newcastle initially covered 870.000 hectares and encompassed over 6.700 properties, mostly privately owned farms and land owned by companies. During the application stage, Rhino Oil and Gas reduced the application area from an initial 3.2 million hectares because of growing pressure from opponents and social movement actors (Erasmus 2016). The company maintained that this downsizing was a part of the common application procedure, while FFSA saw this as a victory of their own mobilisation. For this early stage of the application process, the company was allowed to fly over the application area with a specially equipped plane to survey 'Earth's gravitational field to help image subsurface structures' (Rhino Oil and Gas 2016, p. 7).

Rhino's initial application in Central KZN encompassed the Midlands as well as the southern outskirts of KZN's provincial capital, Pietermaritzburg. The exploration area reaches far into the northeast of the province but excludes a number of protected areas. The applications are overseen by SLR Consulting, which is the same environmental consultancy firm supervising the applications in the Newcastle area where redrafting of the footprint area has been quite substantive. A large part of the application was taken out of the original application, which formerly included Greytown and Pietermaritzburg as well as several protected areas such as the Hluhluwe–iMfolozi Game Reserve.

The Frack Free Leadership Group

FFSA was launched in 2015 as a reaction to fracking applications in KwaZulu-Natal and the neighbouring provinces. The movement opposes fracking on 'ecological, health and economic grounds' (FFSA 2021a). Core group members are particularly concerned about conservation areas and excessive water use in an already water-stressed country. FFSA is an umbrella organisation supported by environmental NGOs, conservancies and local businesses such as restaurants and bars, which have declared themselves frack-free all around the country.

FFSA was founded in the KZN Midlands where the support base mainly resides. The Midlands are an affluent part of the province and is home to high-income earners compared to the national average (see Appendix). The most common land uses include ecotourism, agriculture, and forestry (KZN Top Business 2022). Conservation-minded people in the Midlands support local food initiatives and the local conservancies. Because of a critical mass of conservation-minded people, green organisations like Dargle Conservancy Forum or Midlands Conservancy Forum have quite a strong support base. A number of blogs on local sustainable lifestyle practices are hosted in the Midlands. These blogs discuss how to grow food sustainably, live low-carbon lifestyles, and promote barter markets. Through newsletters and personal networks, people were quickly informed about exploration applications in the Midlands. The initial shock created some momentum with a number of people asking the leadership group how to volunteer their time.

Much like most grassroots environmental groups, the campaign was short on finance. The campaign was running on a shoestring budget, which consists of donations and collaborations with environmentally-minded artists auctioning their work for the cause (FFSA 2021b).

FFSA has no formal organising structure. FFSA's organising practices centre around the idea of flexibility. Everyone who wants to use frack-free material is welcome to use or to start a local chapter. Internal communication is ensured via online newsletters and chatgroups. The activists who started the movement chose not to elect official office bearers. However, while there are no official or elected leaders, two unofficial leaders can be identified who steer the direction of the campaign. The first, Nikki Brighton, is responsible for public relations and serves as an unofficial spokesperson by writing blog posts, engaging with the media and inviting people to public events. In an interview, Nikki specified that she volunteers '80%' of her time for Frack Free and other grassroots environmentalist causes in her neighbourhood (activist interview, 20 January 2017). She is well respected among other activists for her enthusiasm and unwavering commitment. The second, Judy Bell, has 20 years of work experience in environmental impact consulting, particularly with water and waste management. Judy declares that fracking is 'about money . . . it's about nothing else' (activist interview, 24 October 2016). Furthermore, she talks about collusion between government officials and the oil and gas industry driven by 'greed'. Judy knows that environmental advocacy is about persistence, and she reports how she has stepped on the toes of regulators all her professional life (ibid.). Both Nikki and Judy know the ecosystem of environmental professionals and activists in the province. Another instrumental member of FFSA is the award-winning environmental activist Bobby Peek. More than two decades ago, he founded the NGO GroundWork, which fights for energy and climate justice. Over the years, Bobby has also established a public profile that helps him to amplify messages that come from the grassroots. In 2014, GroundWork published an anti-fracking report called *Shell: Don't Frack the Karoo*, which cautioned against the social and environmental risks linked to fracking even before FFSA was founded (GroundWork 2014). The movement base of FFSA comprises people coming from similar professional and social backgrounds. Members of FFSA have professional backgrounds in public relations, environmental science and consulting as well as filmmaking. Many members have a long involvement in conservation work, and a number of them have acquired professional experience in environmental impact assessments. Typically, Frack Free activists are concerned about the water use and procedural soundness of the impact assessment. Many of them had also previously come in contact with issues around fracking prior to the applications in KZN, so it was not entirely new territory. Through exchanges with activists in the Karoo, where fracking has been resisted for a couple of years now, conservationists in the Midlands became aware of the government-backed plan to get fracking ahead in their country. The engagement of individual Frack Free members varies, but the majority of members are white and between 40 and 60 years old. The political direction is sometimes contested among the members of the movement. Quotes like the following taken from an FFSA publication are seen as too radical by other members:

> Frack Free SA is powered by volunteers – including Frackivistas . . . who have drawn a line in the sand and [are] determined to take back the power

from the politicians and greedy corporations who have short-term gains as their objectives and do not care about the havoc left behind when they pocket their money and move on.

(FFSA 2016b, p. 3)

According to some more moderate members of FFSA, statements like these might be too radical and scare off opponents to fracking with more conservative agendas. One activist wonders whether FFSA is inclusive enough:

[I]t can be seen to be an extremist green movement . . . and the more conservative, the more sort of pro-development, the average sort of person just going about their life might shy away from something that seems a bit too out there and a bit too greeny. So, I'd say that is the negative side.

(activist interview, 26 January 2017)

Despite some controversy on the framing of the movement, there was a more general sense among the uniting participants that fracking is a serious threat to the land, including water resources.

Four Tactics Against Fracking

The resistance to fracking converged around four tactics that are discussed in the following section. As we will see, a lot of resources are spent on capacity-building. The movement's tactics include blocking and delaying, educating, connecting, and prefiguring. Despite fracking being a moral shock to the area, the fracking threat consolidated the movement and connected it to the larger community.

Blocking and Delaying

So then the real threat came now. So we now heard that the applications were granted, but then suddenly a meeting came . . . so we heard about it like three weeks before they were coming. So I just got [the] publicity machine rolling – fast! So we were just good at that and we managed to get over 200 people to the first meeting, and they were shocked.

(activist interview, 20 January 2017)

As seen from this quote, when the applications for fracking became public, a group of grassroots environmentalists in the Midlands reacted immediately. Members circulated messages that listed the dates and venues of the meetings, encouraging attendees to 'stand against fracking'. Chatgroups and Facebook groups were created ad-hoc, which shared the dates and venues of the first consultation meetings with Rhino Oil and Gas. The first consultation meetings that were scheduled by FFSA were overflowing with people who came to take an active stance against fracking. Banners in the halls read 'We can't drink gas',

'You might as well kill us all #nofracking', or 'Fracking in the wild will poison your child'. The first episodes of encounters between Rhino Oil and Gas who arrived with their environmental consultants and the public were very contentious, highlighting the initial success of the movement. People from NGOs, the farmers union and local residents came to show their opposition to fracking. At the first consultation meeting of Rhino Oil and Gas in Lions River, a participant asked the crowd to lift their hands if they were in favour of fracking. Among the crowd of around 400 people, nobody lifted their hands. When the reversed question was asked, i.e. who was against fracking, everyone lifted his/her hands. The crowd, some of whom had never seen each other before, shared a moment of recognising each other in their opposition to fracking. Participants managed to stage unequivocal opposition to fracking by drawing a line between the oil and gas company and the participating crowds. A number of these meetings had to be postponed, as the venues hired by Rhino were too small for the number of participants. FFSA, thus, delayed the application procedure, which created additional costs for Rhino. Time and money went into finding new venues and inviting the interested public to the meeting. Additionally, the press picked up on instances when public meetings were interrupted and noted that participants were overwhelmingly in opposition to fracking. A subsequent press article investigated Rhino Oil and Gas and noted a lack of capacity, characterising the company as '[a] one-man show run from a shared office block in Cape Town' (Umraw *et al.* 2015). Other press articles quoted comments from the audience at consultation meetings, such as a farmer likening fracking to Russian roulette in Vryheid (Rodway 2016) or the CEO of the farmers union addressing members and saying, 'register and attend the Public Hearings as a matter of importance, the future of our province is at stake' (Northern Natal News 2015). The favourable news coverage helped to sustain mobilisation at consultation meetings for some time.

Much like other mobilisations, FFSA's campaign against fracking has its ebbs and flows. Gradual demobilisation had to do with the extensive application procedure. Public participation meetings took place so frequently that FFSA was not able to attend every meeting in large numbers. Over time, the blocking tactic was no longer viable, as fewer people attended the numerous fracking meetings. An environmental consultant who supports FFSA criticises this 'incrementalism' for obfuscating what is actually at stake (interview with activist, 19 January 2017). While blocking and contesting the oil and gas companies in public meetings was on the decline after the first months of fierce opposition, the focus became on delaying the application process by submitting formal objections.

This can also be interpreted as a manoeuvre to offer lower-level forms of engagement to activists. Those activists working as environmental consultants compiled detailed guidelines on how to comment in writing on the ongoing application process. FFSA encourages the submission of written comments to create extra work and obstacles for the applicants. Written comments need to be compiled by the applicant and submitted to the Petroleum Agency of South Africa

(PASA). FFSA advises opponents to fracking to engage with the environmental impact assessment in the following ways (FFSA 2016c):

- Continue to submit queries/concerns/requests on the process.
- Be specific.
- Be persistent.
- Note the deadline for the process and submit comments on time.
- Even if the Petroleum Agency approves of an application, there is a time window for appeals.

A number of activists are unsure whether the appeals strategy actually works. As seen in Chapter 3, there are doubts whether public participation is actually fair and inclusive enough. For some activists who observed different application procedures, environmental authorisation becomes more and more like a box-ticking exercise (interview with activist, 19 January 2017). In fact, the incrementalism underlying the application procedure is what activists oppose strongly. There is no room in the procedures to oppose fracking overall but to comment on small technical applications towards possible extraction. Delaying the application procedures is, thus, perceived both as an opportunity and as a risk of losing a number of activists in a long, draining, and technical process.

Educating

FFSA's core group ran an educational campaign aiming to counter the claims and valorise the land considered for fracking. This included written material, film screenings, and visits to schools. While the campaign was targeted at a larger public, understanding fracking was also instrumental for the self-understanding of the group. *Fracking in South Africa: A beginner's guide*, which was published in 2016 by FFSA, engages with the history of fossil fuels in South Africa and debunks some of the claims of the fracking industry. The report is illustrated with a number of pictures of birdlife and scenic imagery of South African mountains and rivers. The report opens with the statement that 'diverse communities' have shown their 'widespread opposition' to fracking and that it 'has no place in South Africa' (FFSA 2016b, p. 3).

FFSA's fracking report examines South Africa's technical, legal, and environmental conditions in relation to potential gas drilling in the country. The following findings are brought up to caution against the industry:

- Fracking risks excessive water use and groundwater contamination in an already water-scarce country.
- The economic potential and job creation potential of the industry are likely overstated by the industry.
- The impacts of fracking on climate change, especially from methane escaping as a result of the production process, have to be considered against the backdrop of fracking gas being promoted as a source of clean energy.

- FFSA suggests that renewable energy in places like the Karoo is a safer investment than gas.
- South African law is not sufficiently protecting against land grabs.
- The report ends with a section on 'What can you do?', including the call to register oneself as an interested and affected party in relation to the environmental process, and to attend consultation meetings and voice opposition in those meetings.

(FFSA 2016b)

The beginner's guide to fracking helps activists to contextualise the plans for gas extraction in South Africa. It both counters claims concerning the economic potentials of fracking and addresses the environmental threats for local livelihoods. The material that FFSA provides is also meant to counter the claims put forward by the oil and gas industry. The industry-funded studies sponsored by Shell and Rhino Oil and Gas primarily look at the economic benefits without analysing the concrete repercussions of drilling for local communities, livestock, and farming.

FFSA's educational initiative is aimed especially at deprived neighbourhoods with low levels of formal education. Part of the initiative also involves organised screenings of the US–South African documentary *The High Cost of Cheap Gas*, which are sometimes followed by Q&A sessions attended by the director Jeffrey Barbee himself. The film portrays the corporate behaviour of the gas industry in the USA, exposing the damage it has done to communities and the potential environmental damage it could cause in South Africa. The beginning of the film suggests that fracking triggered the largest land grab in the history of the USA, especially since the technologies of horizontal drilling and hydraulic fracturing were combined in the late 2000s. In the film, families and farmers talk about the repercussions of the fracking industry and the destruction of their livelihoods. According to the activist who organised the screenings of the film, it had a positive impact on mobilisation (interview with activist, 22 October 2016).

One pillar of the work of FFSA activists is school education. Environmental education is part of the conservation work that some Frack Free members do, mostly on a voluntary basis. During school lectures or on school trips, they address the potential dangers of fracking on water resources, wildlife, and humans. Subsequently, some schoolteachers decided that they wanted their students to get engaged and contributed with kids' drawings against fracking. These drawings were used at a protest, showing that opposition to fracking is an intergenerational concern.

Connecting

Fracking is a very expansive industry as the US example shows. This is a challenge to activists who feel isolated, as like-minded people are often hundreds of miles away. Online activism might be helpful to keep people informed, but it tends to encourage short-term engagement due to a lack of personal bonds. While participants foster movement ties via the web, strong ties are unlikely to emerge from internet-based organising (Diani 2000). Clickivism might sometimes amount to 'slackivism', which is defined as a form of 'self-aggrandising' and 'politically ineffective' form of activism in the digital age (Cabrera *et al*. 2017). For movements where participants cannot meet regularly in the neighbourhood, the risk of demobilisation is very real. In anti-fracking struggles, it is thus key to create places of encounter where activists exchange ideas and build connections. These encounters have educational components but are mostly geared towards a shared feeling of belonging to the same struggle. FFSA leaders are keenly aware of the need to 'connect the dots' between different movements and to show that different applications for hydraulic fracturing are part of the same struggle (interview with activist, 24 October 2016). Essentially, this is also a search for allies and political opportunities. In trying to connect to other struggles, FFSA also learnt where other actors stood on the issue.

In October 2016, the Frack Free Festival took place in the small town of Matatiele in the Eastern Cape to build a more integrated community among anti-fracking activists. FFSA contacted organisations from around the country to join. The festival went on for two days, including workshops and brainstorming sessions for communities exchanging their experiences with the industry, information sessions on environmental impact assessments and sessions on alternative energy from sun, wind, and biogas. The festival brought together environmental NGOs, conservancy groups, local traditional councils, and faith groups. One hundred and twenty participants from 31 organisations around the country were able to learn from each other's struggles against the extractive industries and discuss alternative futures. The power of 'unity' between local, national, and even international struggles was highlighted by the experienced Durban-based activist Desmond D'Sa who has been fighting the petrochemical industry for two and a half decades (Harper 2020). In his address to participants, D'Sa spoke about the importance of movement-building in the face of destructive extraction companies.

> Unity is something that takes time. Environmentalists need to talk openly and have frank discussions. It's very important that all voices are given the opportunity to come together. Even small voices must be heard. Everybody matters. That's how we develop a society. Unity is so important. We need to cross the road that divides us, make friends with neighbors and build trust. . . . Mining is a curse, it is not benefiting anyone, people are dying. Fracking is being claimed as clean energy. That's a lie. Coming together we are part of an international struggle.
>
> It is critical that we continue to build and carry on the struggle and listen to one another and share information. We have a common struggle. It's never just local, it's worldwide. How do we link our local struggle to the international struggle? Solutions lie with us. We know the land and we need to fight for it. Once you kill the soil you can't fix it. Let's not be fooled by these powerful people. They control the world and ordinary people never benefit out of it. Only by working together, will the solutions for the future be found. We can stop these people.

> (FFSA 2016d)

Through exchanges, activists mapped the various issues identified around fracking. The key to fracking resistance for the activists was finding a way to respond to the associated triple challenges around food, water, and energy. Fracking is seen to be in conflict with all these key challenges and political elites and corporate actors seen to curtail communities' say in decision-making. The anti-fracking manifesto that was written and presented in Matatiele blames 'undemocratic and divisive tactics used by corporations and government in our communities to push their version of a development agenda' (FFSA 2016d). One key concern was to find ways of scaling up the protest and strengthen the local voices at the same time.

The invitation to different political stakeholders was answered by few political actors. In order to put pressure on political actors, FFSA had also invited government officials from different levels, traditional authorities, and political parties to the Frack Free Festival. The response rate of parties and government authorities was low. However, Cheryllyn Dudley, a member of parliament from the fringe party African Christian Democratic Party (ACDP) joined the proceedings of the Frack Free Festival. Even though the ACDP is a conservative party that does not share FFSA's mostly left-leaning credentials, the anti-fracking issue unites them. In an interview with Dudley, who became an MP in 1999, she explained to me that the fracking issue raised a red flag with her because she represents constituencies already suffering from water scarcity (interview with member of parliament, 8 February 2017). During the Frack Free Festival, Dudley and FFSA activists agreed that a moratorium on fracking should be the minimal demand to the national government. Towards the end of the proceedings, Dudley promised activists to try and schedule a meeting on fracking in parliament.

Through active lobbying of the ruling party, Dudley managed to initiate a public hearing in parliament only two months later. Dudley opened the debate in parliament by relying on a number of FFSA's arguments from the Matatiele Manifesto. Dudley proclaimed before parliament that 'now more than ever, we need to know exactly who is capturing our water resources and land, and exactly who will benefit' (FFSA 2016a). In her speech, she also questioned the integrity of environmental consultants and refuted the economic case made in favour of fracking.

The parliamentary debate established the political parties' stances on fracking; many of which had not previously spoken on the issue. For activists, the debate was also instrumental in showing that ANC politicians used arguments taken directly from the oil and gas industry. The episode showed that establishing connections to political parties can be useful to bring the topic of fracking risks to the agenda. However, the debate also showed that the ANC was prepared to use the opportunity to advertise the use of unconventional gas. In their line of argument, fracking was directly linked to the national interest. For instance, MP Pikinini of the ANC concluded his speech by demanding that 'the eyes of all patriots should be on what is good for the nation. Therefore, fracking and other means of gas extraction as game-changer is what is good for the nation at this point in time' (FFSA 2016c).

As the Frack Free Festival illustrates, FFSA succeeded at bringing together a number of organisations that spoke up against fracking in South Africa. Despite the initial low response of political parties, reaching out to them paid off eventually, as it led to the debate on fracking in the national parliament. In parliament, the Matatiele Manifesto was mentioned by fracking proponents and opponents alike lifting FFSA to the status of a noticed political player on the issue. Thus, the efforts to connect the dots amplified FFSA's message.

Prefiguring

Applications for fracking are re-territorialising projects. In the application area, existing land uses such as farming, conservation, and residential housing come under threat by fracking applications. One way to counter applications is to strengthen and celebrate frack-free communities. While blocking and delaying is a reactive tactic to external threats, prefiguring aims at deepening communal practices and emphasises and supports sustainable livelihoods. Generally, prefigurative politics in social movement scholarship describes political practices that actualise the changes that activists want to see in the world (Yates 2015; Leach 2013). Attempts of prefigurative networks can be found at all scales from the neighbourhood level where local democracy gets enacted in inclusive circles to the global level scale where the World Social Forum (WSF) builds capacity towards a post-neoliberal world order. Scaling up protest is seen as a necessary albeit difficult task in order to bring attention to the movement and its demands.

Frack Free encourages bystanders and observers to prefigure a prosperous frack-free society in everyday acts and at the neighbourhood level. To this end, FFSA offers different levels of engagement. Fracking resistance starts with very mundane acts. At the individual level, FFSA encourages farms and local businesses to declare 'Frack Free zones' (FFSA 2021a). A number of local businesses also show their opposition to fracking by having their names on their own website as well as the Frack Free website. Some of these businesses paste stickers somewhere in their shop or on their car. Some others display leaflets on their counters. Such acts are less demanding than participating in a demonstration, for example. The bright Frack Free purple and green colours spark conversations about fracking. Declaring a personal 'Frack Free zone' around a farm might inspire others to do the same. Furthermore, Neville and Weinthal (2016) have shown how fracking site disputes can ultimately redefine the meaning of 'the local' and thus bear the potential to be scaled up. According to them, 'regard for one's neighbourhood, community, and local surroundings, as well as concern for one's safety and well-being – may be strategically used to drive broader, more encompassing activism and participation' (ibid., p. 19).

At the neighbourhood level, some of the communal activities are branded frack-free. Barter markets in the Midlands and organised tours of conservancies are advertised on the Frack Free website. Participants of these activities are informed about the environmental risks of fracking for the neighbourhood and the environment.

FFSA also advertises the use of renewable energy and encourages practices such as cooking with the help of solar cookers. Fossil-free lifestyles are a way of walking the talk. Strategically, mundane acts if framed in the right way can be recognised as frack-free acts. This links everyday practices with causes on a larger scale.

Discussion and Conclusion: How to Fight Fracking and Coal?

Social movement mobilisation is shaped by and shapes political space. Fracking is a spatial challenge for protest movements that results from the vast footprint of fracking applications. Initial exploration applications covered 40% of KZN province which caused a moral shock and lead to the formation of FFSA. In the course of the technical application procedure, the fracking applications are usually re-visited and the impact area gets reduced. Some applications by oil and gas companies lapse and other oil and gas companies apply for the same area. FFSA was confronted with a number of different spatial constellations since the first applications shocked the environmentally-minded constituencies mainly in the KZN Midlands. Rhino Oil and Gas alone has applied for seven different exploration rights since 2015. The applications combined are seen as threats to the heritage and environmental balance of the region. In light of these multiple challenges, FFSA had to take strategic decisions to oppose the industry. FFSA relied on several movement tactics both to consolidate the movement internally, but also to make public appearances in confrontation with their adversary.

Thus, in the case of anti-fossil-fuel mobilisations, social movement leaders have to pay particular attention to contested and changing geographical constellations. The fracking application procedure is similar to a cat-and-mouse game bearing a lot of risks for de-mobilisation. Activists mobilise to oppose an application that might be re-drafted or overall dropped soon after. The threat of fracking appears on the scene in one place to re-emerge somewhere else. Core activists lament that the process is demobilising because of the frequent and technical consultation processes. In order to follow the application trail, activists have to travel long distances to arrive at the spaced-out town hall meetings. The meetings are very technical, and the oil and gas company is unwilling to talk about the overall prospect of fracking. Every meeting is scheduled for very specific operations that the company applied for. As the meetings are always conducted by the company that also structures the order of the day, activists have to decide how they intervene in the process strategically. If left unchallenged, the way in which the oil and gas companies present the applications to interested audiences risks convincing a critical mass of the drilling technique.

Timing plays another challenging role for social movement activists against fracking applications. The period between exploration and extraction is likely to be between 10 and 15 years if the applicant receives a production right from the government. The fracking application schedule is opaque even for activists who are familiar with environmental impact assessments (EIAs). From the point of view of social movements, tactics thus need to increase the likelihood for the movement to survive over a longer period. Dormancy is not necessarily a problem if leaders manage to 'remind' the public from time to time that the movement continues with their activism. A number of FFSA's tactics aim at enlarging the movement base. Public appearances at consultation meetings or at marches are only the tip of the iceberg of FFSA's activities.

In line with previous work on social movement tactics, this chapter showed that 'tactics are not only externally oriented but play important roles in movement organisations, such as building solidarity among participants' (Smithey 2009, p. 661). This chapter showed that in addition to blocking and delaying tactics, FFSA uses a host of resources on internal movement-building such as educating, connecting, and prefiguring.

For FFSA, it was possible to mobilise large crowds as a reaction to the moral outcry caused by the fracking applications. The organisation and channelling of outrage is a key to social movement success. Protests were reported on in the press so that a larger audience learnt about the potential threats of fracking. These initial public appearances that were closely identified with FFSA as a movement ensured that the movement was able to mobilise lasting support.

In these public confrontations, FFSA was able to win a symbolic victory by rendering visible the opposition between those in favour of fracking and those who stand against it. By polarising the issue of fracking, FFSA actively sought to bring the wider public on their side.

Education tactics are important to raise awareness, as many people have not heard about fracking before. Whether the people will become proponents or opponents of fracking will also depend on whether they hear from fracking from Rhino Oil and Gas or FFSA first. While the information strategy of Rhino Oil and Gas is very piecemeal, FFSA tries to give a more holistic picture of the negative impacts of fracking. Crucially, FFSA objects to the narrative of fracking as a 'game-changer' thus putting in doubt claims on job creation and local economic development. Both the screening of films and visiting school classes were used to reach audiences that would typically not hear about energy politics otherwise.

Connecting to other groups was particularly successful when FFSA organised a festival inviting a host of different organisations from around the country. However, connecting with political parties and representatives of relevant ministries was tough. FFSA found an unlikely ally in Cheryllyn Dudley, a member of the conservative African Democratic Christian Party (ADCP) who managed to table a parliamentary debate on fracking only two months after the Frack Free Festival. Even though the ruling ANC used the debate to make the case for fracking in the country, there were a number of MPs from other parties raising concerns about South Africa's preparedness for future fracking.

Prefigurative politics helped to re-package some of the everyday acts of activists as frack-free activities. For example, declaring 'Frack-Free zones' effectively means nothing more than labelling everyday practices frack-free. Other mundane acts such as cooking without the uses of oil and gas feature on the FFSA webpage as commendable practices.

Let me now compare the two different movements and flesh out some commonalities. Despite the two different approaches taken to study the

SAVE and FFSA campaigns, some similarities are apparent. Fossil-free movements gain momentum from moral shocks that are translated into political mobilisation.

Sustained opposition to fossil-fuel projects in South Africa challenges deeply entrenched interests within the state apparatus. Both the South African anti-fracking movement and the anti-coal mobilisations challenge state-backed development projects for fossil-fuel extraction. In the process of mobilisation, both movement campaigns learnt about their adversary and underwent internal learning processes. Social movement mobilisation against fossil fuels is very demanding. As we have seen in Chapter 3, the extraction of abundant domestic supply of coal sustained successive political regimes. Social forces confronting fossil fuels in South Africa today are thus fighting a historically almost unchallenged regime. Arguments around affordability and energy autonomy remain persuasive considering especially the rampant unemployment rate in South Africa. Therefore, resistance movements to fossil-fuel projects are facing some limits in the political structure of South Africa.

Political opportunities are not ready-made political spaces ready to penetrate. Rather they need to be created by the movements themselves. In contestation, social movements learn how to make strategic choices in terms of framing and tactics. From the earlier case studies on SAVE and FFSA, some more general lessons can be drawn for our understanding of fossil-free movements. I will draw some final conclusions regarding framing processes, the implications of space for the analysis of social movement analysis and movement tactics.

There are at least three lessons from the two preceding chapters on anti-fossil mobilisation. First, *framing processes* around resistance to fossil-fuel infrastructure do not necessarily start from the explicit opposition to fossil fuels as such. As the case of the SAVE mobilisation showed, frames are negotiable, malleable, and responding to *ad hoc* challenges. FFSA managed to shift and extent its frames at crucial political junctures. The opposition at Fuleni started from grievances of conservationists who framed the mining threat in the terms already familiar to them. The SAVE campaign first defended the conservation park and drew on their leaders' long-standing expertise in conservation. By rallying around the conservation frame, however, the leadership group faced serious limitations for mobilisation. The conservation frame is historically tainted as evictions have repeatedly taken place to establish the park in the first place and enlarge it later on. These past evictions are still inscribed in today's collective memory of residents around the park. The particular cultural script the conservation frame represents ties into a long history of racial injustices concerning land use in Fuleni. The conservation park as we have learnt earlier is still referred to as 'boundary maker' (Brooks 2008, pp. 295–296). The resonance to this frame was impressive internationally but had little resonance with residents around the park. Until conservation practices in South Africa might become more inclusive benefiting society at large including the poor, conservation will not find resonance for social movement mobilisation.

Frame shifts can be effective in mobilising new activists and audiences. By shifting the frame to demands coming from local frontline communities, the campaign managed to enlarge its followership. The campaign's leadership team benefitted from the inclusion of a new member who was able to articulate grassroots demands. The involvement of experienced socio-environmental activist Sifiso Dladla was a key in supporting more confrontational movement tactics. Similar to the FFSA campaign, SAVE used public participation meetings to voice discontent through the application procedure. Particularly the apparent disrespect of Rhino Oil and Gas caused outrage among local residents. Blocking public participation meetings was a symbolic victory for the SAVE campaign.

Particularly, the structure of rural politics inhibited some residents to participate in campaigns against mining. Traditional rural decision-making structures are often revolving around the chief's will. The chief in Fuleni was unhappy with locals speaking up against the mine and thus made opposition to mining an issue of disrespect for traditional structures. Subsequently, there were both splits between supporters and opponents of mining *as well as* between loyalists and opponents of traditional politics and the chief. Mining politics thus proved to be highly contentious in Fuleni.

The death of an anti-mining activist opened an opportunity for the movement to reformulate their demands in line with broader issues such as the destructiveness of extractive industries and the opposition to fossil fuels. The active use of the EJ framing widened the scope of the campaign. While EJ framing alone cannot resolve local tensions around grassroots politics, the frame provides a structural analysis of injustices of large polluting infrastructures. Demands for EJ are frequently used among NGO actors and progressive academics. With some NGO actors and academics taking ownership of the cause, the SAVE campaign became more frequently used in NGO reports and academic papers.

EJ connects racial, social, and ecological demands including nature conservation and the social demands of local residents. When the EJ frame was embraced, the expansion of fossil-fuel infrastructure was constructed as a problem shared between communities in different localities around South Africa and beyond. As we will see in the final chapter on global perspectives, the EJ framing has the highest potential for cross-border mobilisation.

To sum up the conclusions on framing extractive protests, I observed that the framing of resistance relied on leaders' flexibility and understanding of the political moment. Leaders' ability to seize the moment is crucial for SAVE to evolve. Obviously, there is no way of knowing what would have happened had SAVE continued as a conservation campaign. It seems quite likely, however, that ties to frontline communities as well as NGOs and progressive academics would be infrequent at most. More likely, the SAVE campaign would still be frustrated by the political inaction of the conservation community that is historically rather conservative. As a wildlife campaign, SAVE would also be a nonstarter for environmental grassroots organisations such as some NGOs that now stand in solidarity with SAVE.

Second, social movement tactics are geared both to increase the numbers internally and to challenge the adversary. Fossil-fuel politics in South Africa is characterised by very uneven power relations as the state streamlined participation procedures and depoliticised highly contentious mining projects by framing them as development. In this context, it is difficult for social movement actors to raise awareness for the social and environmental ills linked to mining. Movement tactics are particularly effective when movements are able to mark a strong frontier between their own livelihoods and mining projects. If movements are able to show that fossil-fuel projects are in contradiction with a particular way of life, the likelihood of support increases.

FFSA chose a mix of institutional and extra-institutional politics to make claims against the prospect of fracking. Blocking and delaying public participation meetings disrupted institutionalised politics. To the interested public, these encounters bring to light the arguments against mining applications. However, FFSA was only able to make public appearances as long as the public outcry against fossil fuels mobilised large crowds. Successive consultation meetings are part of a long and tedious string of meetings.

The third lesson from this study is that the location of the fossil-fuel projects matters. Fossil-fuel projects are inserted in particular space histories that have an impact on the way people will perceive fossil-fuel projects. The location of fossil-fuel projects under application cannot be disentangled from the understanding of mobilisations.

Conflict dynamics differ depending on the size and location of the impact area. Anti-fracking mobilisations have to find tactics that bring together the residents of farms and residential areas that might have little in common but the common threat of fracking mining. Coal applications on the other hand are for a confined impact area. Both applications for coal and fracking share that they do have a profound influence on local livelihoods.

The spatial impact of fracking applications bear both potentials and risks for social movements. The large footprint of fracking applications potentially increases the number of fracking opponents. On the other hand, the vast spatial impact is a danger as people are physically disconnected from each other. Connecting a dispersed group of affiliated people makes mobilisation difficult. Creating a common 'we' in places such as the Frack Free Festival has resulted in a strategic alliance between a member of parliament and Frack Free activists.

The dominant story of any land is difficult for anyone to change. If social movements find a compelling argument that links historical land-use practices to the integrity and inherent value of the land, mining companies will need to make very convincing arguments in favour of mining. Job creation and economic development are often being invoked by mining companies that are very aware of poverty and high rates of unemployment in rural South Africa. Additionally, mining companies have a tendency to downplay the impacts of their operations. Thus, in places that are seen of particular value to social movement actors, fossil-fuel hegemony is at its weakest.

South Africa's modern industrialising political regimes have relied on the extraction of coal. This deep structural dependence has naturalised coal mining and the use of domestic fossil-fuel supply mostly for cheap energy, but also for export revenues. So far, the South African mining regime was challenged by labour unions for better pay and better working conditions. The current opposition to mining questions mining practices more fundamentally on environmental and social grounds. Growing discontent by local communities and environmental NGOs with the existing regime opens up opportunities for social movement intervention. Social movements use a number of tactics both in confrontation with corporate adversaries and to raise awareness for the damaging effects of fossil-fuel use.

Public participation meetings afford opportunities for opponents of fossil-fuel projects to voice public opposition. The initial tactic of FFSA was to stage public protests at public participation meetings in order to disrupt them. These contentious interactions brought public awareness to the cause. A number of newspapers and other media reported about the opposition to fracking. However, the technical nature of the participation process led to the gradual demobilisation of the movement. Subsequently, FFSA focused more on encouraging written objections to the process. The underlying argument was to unmask the technical participation process as a means to implement polluting fracking.

As in the US context, anti-fracking educational practices played a key role in shaping local opinions (Vasi *et al.* 2015). Written and film materials are illustrative means to shape public opinion and share information widely. In the fight against fracking in South Africa, the documentary *High Cost of Cheap Gas* was a means to raise public awareness. Flashing out global connections between the US context and South Africa, the documentary manages to make a strong case against fracking in South Africa. Also, the *Beginner's Guide to Fracking* disseminates information on the social and environmental costs of fracking and encourages readers to engage in opposition.

Connecting the movement brings in new perspectives and strengthens ties between social formations that often had no prior contact. From a strategic point of view, a crucial question is whether the anti-fossil-fuel agenda can be linked to other political demands that allow the movement to formulate a more comprehensive political agenda. A post-extractivist agenda would need to be linked to other social and economic demands. Influential civil society actors such as labour unions would need to get convinced of a post-extractivist agenda as they still form the largest civil society organisations in South Africa. To find common ground between a post-extractivist agenda and labour unions that organise mine workers seems highly unlikely. Labour unions are interested in raising wages, shortening the working day and improve safety measures. For unions to attack the mining industry itself and calling to close the mines would be the end of them. Social movements would therefore need to start a conversation on a comprehensive post-extractivist transition. This conversation is extremely difficult and requires the intellectual leadership of 'organic intellectuals' that are able to connect the dots between politics, industry, and civil society.

Finally, the promotion of alternative post-extractivist lifestyles and active de-linking from fossil-fuel-intensive practices can be regarded as prefigurative movement practices. Prefiguration empowers social movements as they start to act on the kinds of ideas they want to see in the world. Declaring a business or farm 'frack free' suggests to consumers that fracking is not inevitable. These everyday acts have the potential to bring into conversation people living in a neighbourhood about potential dangers to their way of life. As difficult as this conversation might be, in the preceding chapters, we have seen that fossil-fuel hegemony is more and more contested in local neighbourhoods. These movement campaigns from below aim to prefigure a politics that would need to be scaled up.

The final chapter is on global perspectives against fossil-free mobilisation. The chapter first deals with anti-coal mobilisation in Germany. As a major fossil-fuel emitter in Europe and coal producer, social mobilisation became more and more contested over the last years. The chapter discusses frames and tactics used by the German Ende Gelände movement and draws some conclusions regarding possible convergences between South African and German fossil-free movements. In the last section of the chapter, I discuss the affordances and limitations of different movement frames for a global post-extractivist agenda.

Note

1 It might be worth mentioning that the fracking boom in the US was initiated by smaller and medium-sized 'wildcatter' firms and not the oil and gas majors that were, at first, quite reluctant to bank on fracking (Bradshaw and Boersma 2020, p. 84). In South Africa, there are a number of bigger and smaller companies that showed their interest in potentially contributing to the fracking oil and gas market. The South African energy and chemical company Sasol is investing in fracking in Australia and has indicated its preparedness to start fracking in South Africa 'if it could be done in an environmentally responsible manner and within a regulatory framework' (Sasol 2011, p. 21).

References

Academy of Science of South Africa, 2016. *South Africa's technical readiness to support the shale gas industry*. Available from: https://research.assaf.org.za/bitstream/handle/20.500.11911/14/2016_assaf_sa_technical_readiness_shale_gas_final_report.pdf?sequence=14&isAllowed=y

African National Congress, 2019. *Let's grow South Africa together. 2019 Election manifesto*. Available from: https://cisp.cachefly.net/assets/articles/attachments/77065_6140_anc_manifesto_booklet_a5_digital.pdf

African News Agency Reporter, 2017, July 28. *Fracking exploration firm withdraws KZN application*. Available from: www.iol.co.za/news/south-africa/kwazulunatal/fracking-exploration-firm-withdraws-kzn-application-10528968

Alexander, P., 2010. Rebellion of the poor: South Africa's service delivery protests – a preliminary analysis. *Review of African Political Economy*, 37(123), 25–40. doi:10.1080/03056241003637870

Atkinson, D., 2018. Fracking in a fractured environment: shale gas mining and institutional dynamics in South Africa's young democracy. *Extractive Industries and Society*, 5 (4), 441–452. doi:10.1016/j.exis.2018.09.013

Azubuike, S., *et al.*, 2018. Identifying policy and legal issues for shale gas development in Algeria: a SWOT analysis. *Extractive Industries and Society*, 5 (4), 469–480. doi:10.1016/j.exis.2018.10.005

Bond, P., 1998. Privatisation, participation and protest in the restructuring of municipal services. *Urban Forum*, 9 (1), 37–75. doi:10.1007/BF03033129

Bradshaw, M. and Boersma, T., 2020. *Natural gas*. Cambridge: Polity Press.

Brooks, S., 2008. Royal precedents and landscape midwives: claiming the Zululand wilderness. *In*: B. Carton, J. Laband and J. Sithole, eds., *Zulu identities: being Zulu, past and present*. Pietermaritzburg, South Africa: University of KwaZulu-Natal Press, 293–303.

Cabrera, N., *et al.*, 2017. Activism or slacktivism? The potential and pitfalls of social media in contemporary student activism. *Diversity in Higher Education*, 10 (4), 400–415. doi:10.1037/dhe0000061

Collier, P., 2010. The political economy of natural resources. *Social Research*, 77 (4), 1105–1132.

Diani, M., 2000. Social movement networks virtual and real. *Information, Communication & Society*, 3 (3), 386–401. doi:10.1080/13691180051033333

du Plessis, W., 2015. Regulation of hydraulic fracturing in South Africa: a project life-cycle approach. *Potchefstroom Electronic Law Journal*, 18 (5), 1441–1478. doi:10.4314/pelj.v18i5.06

Energy Information Administration, 2015. *World shale resource assessments*. Available from: www.eia.gov/analysis/studies/worldshalegas

Environmental Justice Atlas, n.d. *Fracking frenzy*. Available from: https://ejatlas.org/featured/fracking-frenzy

Erasmus, J., 2016, February 24. Anti-fracking lobby slowly gaining momentum in KZN. *News24*. Available from: www.news24.com/SouthAfrica/News/anti-fracking-lobby-slowly-gaining-momentum-in-kzn-20160223

FFSA, 2016a. *Frack free goes parliament*. Available from: https://frackfreesa.org.za/index.php/2016/12/08/fracking-goes-to-parliament/

FFSA, 2016b. *Fracking in South Africa: a beginner's guide*. Frack Free South Africa. Available from: https://view.publitas.com/wild-world-books/fracking-in-sa-a-beginners-guide/page/1

FFSA, 2016c. *EIAs made easy*. Available from: https://frackfreesa.org.za/index.php/2016/07/07/eias-made-easy/

FFSA, 2021a. *About*. Available from: https://frackfreesa.org.za

FFSA, 2021b. *Art for the earth*. Available from: https://frackfreesa.org.za/index.php/what-can-i-do-2/art-for-the-earth/art-gallery/

Finkeldey, J., 2018. Unconventionally contentious: Frack Free South Africa's challenge to the oil and gas industry. *The Extractive Industries and Society*, 5 (4), 461–468. doi:10.1016/j.exis.2018.08.006

Frack Free South Africa, 2016d. *Matatiele manifesto*. Available from: https://frackfreesa.org.za/index.php/2016/11/10/matatiele-manifesto/

Fusco, L.F. and Carter, A.V., 2017. Toward an anti-fracking mobilization toolkit: ten practices from Western Newfoundland's campaign. *Interface: A Journal for and About Social Movements*, 9 (2), 276–299.

GroundWork, 2014. *Shell: don't frack the karoo*. Available from: www.groundwork.org.za/specialreports/2014%20August%20NL%20shell%20report%20eng%20rs.pdf

Hadad, M.G., *et al.*, 2021. Socio-territorial disputes and violence on fracking land in Vaca Muerte, Argentina. *Latin American Perspectives*, 236 (48), 63–83. doi:10.1177/00945 82X20975009

Harper, P., 2020, September 13. Q&A sessions: 'nobody will be able to stop us'. *Mail & Guardian*. Available from: https://mg.co.za/environment/2020-09-13-qa-sessions-nobody-will-be-able-to-stop-us-desmond-dsa/

Kinetiko, 2013. *Coal bed methane in South Africa*. Available from: www.kinetiko.com.au

KZN Top Business, 2022. *uMngeni Municipality*. Available from: https://www.kzntopbusiness. co.za/site/umngeni

Ladd, A., ed., 2018. *Fractured communities: risks, impacts, and protest against hydraulic fracturing in U.S. shale regions*. New Brunswick, NJ: Rutgers University Press.

Leach, D.K., 2013. Prefigurative politics. *In*: D.A. Snow, D. Della Porta, B. Klandermans and D. McAdam, eds., *The Wiley-Blackwell encyclopedia of social and political movements*. Hoboken, 1–2.

Milhaylov, N.L., 2020. From victims to citizens: emerging activist identities in the anti-fracking movement in Bulgaria. *Journal of Community Psychology*, 48 (2), 170–191. doi:10.1002/jcop.22258

Munci, E., 2019. 'Peaceful protesters' and 'dangerous criminals': the framing and reframing of anti-fracking activists in the UK. *Social Movement Studies*, 19 (4), 464–481. doi: 10.1080/14742837.2019.1708309

Neville, K.J. and Weinthal, E., 2016. Scaling up site disputes: Strategies to redefine 'local' in the fight against fracking. *Environmental Politics*, 25 (4), 569–592. doi:10.1080/096 44016.2016.1154124

Northern Natal News, 2015, November 5. Fracking threat for Northern KZN farmers. *Northern Natal News*. Available from: https://northernnatalnews.co.za/38500/ fracking-threat-for-northern-kzn-farmers/

Nyberg, C., *et al.*, 2018. Dash for gas: climate change, hegemony and the scalar politics of fracking in the UK. *British Academy of Management*, 29, 235–251. doi:10.1111/1467-8551.12291

Rhino Oil & Gas, 2016, November. *Environmental impact assessment for exploration: right application for petroleum on various farms in Northern KwaZulu-Natal*. Presentation at EIA meeting (PowerPoint slides).

Rhino Resources, 2018. *Areas*. Available from: www.rhinoresourcesltd.com/areas

Rodway, E., 2016, March 17. *Fracking vs farming*. Available from: http://vryheidherald. co.za/40166/fracking-pic/

SASOL, 2011, June 30. *Pursuing responsible growth*. Sustainable Development Report. Available from: www.sasol.com/sites/default/files/financial_reports/Sustainable%20 development%20report%20-%202011.pdf

Scholes, R., Lochner, P., Schreiner, G., Snyman-Van der Walt, L. and De Jager, M., eds., 2016. *Shale gas development in the central Karoo: a scientific assessment of the opportunities and risks*. Stellenbosch, South Africa: CSIR.

Smithey, L.A., 2009. Social movement strategy, tactics, and collective identity. *Sociology Compass*, 3 (4), 658–671. doi:10.1111/j.1751-9020.2009.00218.x

Temper, L., 2019. Blocking pipelines, unsettling environmental justice: from rights of nature to responsibility to territory. *Local Environment*, 24 (2), 94–112. doi:10.1080/13 549839.2018.1536698

Umraw, A., Mthalane, V. and Erasmus, J., 2015, October 19. *Dubious gas rights deal*. Available from: www.news24.com/SouthAfrica/News/Dubious-gas-rights-deal-20151019

Vasi, B., *et al.*, 2015. 'No fracking way!' Documentary film, discursive opportunity, and local opposition against hydraulic fracturing in the United States, 2010–2013. *American Sociological Review*, 80 (5), 934–959. doi:10.1177/0003122415598534

Vesalon, L. and Crețan, R., 2015. 'We are not the Wild West': anti-fracking protests in Romania. *Environmental Politics*, 24 (2), 288–307. doi:10.1080/09644016.2014.1000639

Wasserman, H., Chuma, W. and Bosch, T., 2018. Print media coverage of service delivery protests in South Africa: a content analysis. *African Studies*, 77 (1), 145–156. doi:10.1080/00020184.2018.1426312

Yates, L., 2015. Rethinking prefiguration: alternatives, micropolitics and goals in social movements. *Social Movement Studies*, 14 (1), 1–21. doi:10.1080/14742837.2013.870883

7 Fighting Fossil Fuels Around the World

This chapter outlines comparative and global perspectives on social movements against fossil-fuel extraction. The South African cases analysed in Chapters 5 and 6 will be compared to German social movement action against coal. Here, I show that fossil fuels are entrenched in the German case, but that the state has conceded to some of the demands from anti-fossil-fuel mobilisations. I focus my analysis on a social movement called Ende Gelände that is rallying for an acceleration of the agreed coal exit in Germany. Second, I analyse the potentials of different anti-fossil fuel frames regarding their potentials for global mobilisation. Discussing three different frames, I show that the EJ frame offers the most potential for concerted claim-making between anti-fossil movements from the Global South and the Global North.

Germany: 'Coal Exit Is a Handicraft'

Social movement action in Germany in the past years has more and more focused on 'coal exit'. This chapter discusses social movement struggles against coal in Germany as the biggest emitter from fossil fuels in the European Union (EU) and the seventh largest CO_2 emitter globally (Statista 2021). Seven among the ten largest single sources of emissions in Europe are coal-fired power stations in Germany (Suhr 2019). The German Federal Institute for Geosciences and Natural Resources (Bundesanstalt für Geowissenschaften und Rohstoffe) calls Germany 'an important mining country' only second to China in lignite production ('Braunkohle' in German) (2020, p. 7). Lignite emits more CO_2 than other types of coal (Kahya 2014). There are still three coal mining areas in Germany – namely in Rhineland, Central Germany, and Lusatia (Bundesverband Braunkohle n.d.). Current mobilisation also takes place against the backdrop of the German 'coal compromise', a multi-stakeholder arrangement between the state, energy corporations, scientists, and civil society actors that made policy recommendations. These recommendations finally lead to the coal-phase out act of 2020 that agreed to end coal for energy generation by 2038. German climate movements reacted to the compromise by mobilising for demonstrations as the 2038 target is seen as incompatible with the government's emission targets (Meisner 2019).

DOI: 10.4324/9781003110835-9

In this chapter, I present the German fossil fuel resistance and later compare it to findings from the previous analysis of South African social movement mobilisation against fossil fuels. Despite struggles against fossil fuels being mostly local in nature, local protests against fossil fuels belong to the same 'class of conflict that appear regularly in the world' (Martinez-Alier *et al.* 2016, p. 747). Increasingly we see that social movement actors recognise themselves as part of the same struggle, thus creating the potential for a more globalised fossil-free movement. As we will see, social movements in South Africa and Germany are part of the same 'class of conflict' fighting back the fossil-fuel frontier with sometimes similar frames and tactics. Social movement mobilisation primarily responds to national or local political opportunity structures, but some movements respond to global problems (Tilly and Tarrow 2015, p. 20). However, environmental problems come in different forms around the world and are thus hard to frame in a common language (Uekötter 2014, p. 103).

The remainder of this chapter proceeds as follows. First, the general context of coal extractivism and the role of the German state are discussed. Even before the German reunification in 1990, coal mining played a central role in both German states. Second, I describe the context of the eastern traditional coal mining area Lusatia which became an intense site of anti-fossil fuel protest since 2015. Third, I compare and contrast the anti-fossil-fuel movements in Germany with the ones in South Africa to offer an international comparative perspective on fossil-free movements in major fossil-fuel-producing countries. I conclude with an outlook on the global fossil-free movement with a particular emphasis on North-South social movement relations. I argue that framing around climate justice has more potential for global claim-making than calls for a Green New Deal or de-growth that are embraced by social movement actors (Table 7.1).

The German State(s) and Coal Mining

In post-war Germany's both eastern and western parts, energy policy was almost synonymous with coal production and use (Renn and Marshall 2016). Both German states relied on their domestic hard coal and lignite deposits for post-world war reconstruction. At crucial political junctures, both German administrations decided to entrench coal power. In the German Democratic Republic (GDR) after the oil shock of 1973, the socialist unity party intensified its lignite production which remained the primary energy source until the end of Cold War (Friedrich-Ebert-Stiftung 1988). Coal workers and engineers enjoyed great appreciation by the GDR symbolised by the yearly Miners' Day and were rewarded with extra pay (Müller 2017, p. 218). After German unification, outgoing chancellor Helmut Kohl inaugurated the coal-fired power plant Schwarze Pumpe in 1998 calling it a 'business for the 21st century' in parliament (Deutscher Bundestag 1998, p. 23059). Even though environmental destructions such as acid rain and forest death ('Waldsterben') from coal-burning were problems identified early from the early 1980s (Uekötter 2014, p. 114), coal as a source of energy was not questioned by any significant political force until 2015.

German extraction regulation promotes 'economical and careful use of land' (Umweltbundesamt 2020). In this spirit, a pro-coal post-unification charm offensive brought investment in landscape rehabilitation of former coal mining areas to make up for the dirty image of coal – to great costs and mostly limited success (Renn and Marshall 2016, p. 228). Even the first government between the Social Democratic Party of Germany (SPD) and Greens (1998–2005) remained committed to coal use and instead actively promoted the phasing-out of nuclear energy (Illing 2016, Chapter 8). The SPD remained committed to coal until 2020 when they agreed with their former coalition partner Christian Democratic Party (CDU) to end coal by 2038. The coal phase-out act was adopted after sustained pressure by social movements and national media attention on climate change and coal. The act also agrees to compensate workers over the age of 58 affected by decommissioning of mines or coal-fired power stations (Bundesregierung 2020). While the government celebrated the act as landslide success, anti-fossil-fuel movements were everything but convinced by what was achieved. In fact, movements were quick in pointing out that the act was a wholly inadequate response and that continuing coal for almost two decades would not be in line with emission targets (Fridays for Future n.d.).

The coal phase-out act was accompanied by major concessions to energy companies, which can be explained by the long-term involvement of lobbyists in government. As reported in the press, the government agrees to pay 4.35 billion Euros in compensation to energy companies Leag and RWE (Pinzler 2020). German fossil-fuel lobbyists from major energy companies are influencing political decision making in the energy sector. They slow green transition ambitions that were first initiated by the first coalition government coalition between the SPD and Green party, by toning down or deny the results of climate science (Götze and Joeres 2020). As Götze and Joeres show, there are a number of cases of revolving doors between political elites and the fossil-fuel industry.[1] Fossil fuel subsidies for coal between 1970 and 2012 amount to 400 billion Euros in Germany (ibid., p. 165). Even though Germany is seen as the role model for clean energy transitions, the pace of change is lagging behind the government's climate targets and significant concessions have been made to fossil-fuel interests.

Pro-lignite organisations are also struggling to push the government back to reverse the coal phase-out. There are a number of pro-fossil fuel advocacy networks pointing to the high costs of renewable energy and the loss of jobs in the coal industry. The federal association for lignite in Germany claims that domestic coal is the 'only domestic energy source that is available in sufficient quantities and at competitive conditions. Lignite ensures the security of supply in Germany – even when the wind is not blowing and the sun is not shining' (Bundesverband Braunkohle, n.d.).

Ende Gelände and the Fossil-Fuel Frontier

The current anti-coal mobilisation in Germany draws on some of the practices and traditions of the anti-nuclear movement, alter-globalisation movement, and

British climate camp movements (Sander 2017; Brown *et al.* 2018). The movement can be seen as part of an emerging climate justice movement that demands global socio-environmental transformation (Sanders 2017). A formative event was the 2007 G8 summit in Heiligendamm that inspired the organisation of a number of climate protest camps. The British tradition of climate camps was first taken up in Germany at the 'climate and antiracism camp' in Hamburg in 2008 (Sander 2017, p. 27). Another crucial event was environmentalists' deception of the Copenhagen Climate Change summit outcomes in 2009 (Häußerman and Wollny 2017, p. 34). Ende Gelände's main areas of protest actions are the Rhineland in western Germany and Lusatia in the eastern part. While the majority of people in Rhineland are in favour of exiting coal, in Lusatia coal exit is more contested (Rinscheid 2018). In Rhineland, people are more prone to link coal mining to dirt, air pollution and destruction of the environment, people in Lusatia associate coal mining primarily with jobs but also to smoke stakes and pollution (Rinscheid 2018, pp. 19–20). Even in villages that are being evacuated for coal mining, not every villager took a stance against mining (Müller 2017, p. 217). Müller explains why coal mining has an ambiguous legacy in Lusatia:

> Opencast mines and power plants are omnipresent; they are visible signs of coal mining in everyday life. They can certainly have connotations of destruction and decay, but they are also a sign of the region's economic prosperity. Lignite mining, with its indirect and directly visible and tangible consequences, has long been part of Lusatia and has become part of the region's identity.
>
> (Müller 2017, p. 218, my translation)

The following analysis draws on the experience of Ende Gelände in Lusatia. I discuss the framing the movement employs to mobilise for protests and also look at movement tactics.

Ende Gelände is a social movement that can be seen as part of a new environmental protest cycle in Germany. This protest cycle includes mass mobilisation of the Fridays for Future movement in Germany peaking in 2019 when 1.4 million people were in the streets to demonstrate for a better climate policy on 20 September in 575 German cities (Fridays for Future 2019). 2019 was also the most active year of Ende Gelände so far as they realised two blockades of coal infrastructure in Rhineland and Lusatia (Ende Gelände 2019). Ende Gelände can be seen as the radical fringe of a broad climate alliance in Germany which also includes environmental NGOs and pressure groups. Berlin Intelligence service declared Ende Gelände in Berlin part of the 'extremist left' (Verfassungsschutz Berlin 2019, pp. 162–165). Subsequently, more than 20 NGOs and youth wings of three different political parties declared their solidarity in support of Ende Gelände (Ende Gelände 2020a). Ende Gelände is a left-wing movement that formed in 2015 to accelerate the German coal exit primarily by blocking coal infrastructure in Germany. Some Ende Gelände activists have been part of the anti-nuclear movement or other environmental groups, while some others had no prior activist experience. The majority of activists engaged in mass action and spokespeople of the movement are students living in German cities (Stokowski 2019). The group advocates 'feminist leadership' (Stephens 2020) and aims for more inclusion of ethnic minorities.

Ende Gelände frames its message in decidedly anti-capitalist terms combining anti-coal mobilisation with other tropes of the left including demands for open borders, anti-colonialism, anti-racism, and feminism. The use of coal is also seen as fundamentally undemocratic as the true cost of climate change-related costs from coal is seen not to be taken sufficiently into consideration by the German government. The movement accuses the government to ignore the consequences of climate change to the detriment of people living in the Global South suffering from climate change as well as the local damages for residents living in coal mining areas.

> We as Ende Gelände call for an immediate exit from coal and a socially acceptable transformation of all fossil industries. We want a democratic and decentralised energy transition in which people can decide for themselves about consumption and production. A profound, socio-ecological change is necessary in order to achieve a good life for everyone. We believe that overcoming global capitalism, its growth pressures and mechanisms of exploitation are essential. We do not believe that climate change will be stopped within this capitalist economic system.
>
> (Ende Gelände 2017)

A common feature of the new generation of climate movements in Germany is that it links its claim-making to the 1.5-degree goal as agreed in the Paris Climate agreement. Ende Gelände demands an immediate exit from coal, while the Fridays for Future movement demands a coal exit until 2030 (Fridays for Future n.d.). Both movements share that they criticise the German government's inertia in addressing coal phase-out. Fridays for Future has also been more reluctant to link coal-exit to a critique of capitalist accumulation strategies and primarily insist that climate and energy policy should be in line with climate science. While Fridays for Future primarily mobilises schoolchildren 'on climate strike' in urban centres, Ende Gelände's primarily tactic aims at blocking of coal infrastructure, as I discuss more in the following section. Analogously to the last chapter, I also discuss educating, connecting, and prefiguring as social movement practices against fossil fuels in Germany.

Blocking and Delaying

Ende Gelände's main tactic is to block coal infrastructure. Already in 2011 a small group of activists had tried to block a coal mine in the Rhineland but were unable to make a lasting impact. The first action under the banner of Ende Gelände was prepared and realised in 2015. Preparation meetings where direct action was practised took place in different cities around Germany such as Cologne and

Hanover (Ende Gelände 2015). Related activities such as a climate camp and a Degrowth Summer School were also advertised and attended by Ende Gelände activists in preparation to the blockade (Ende Gelände 2015). Finally, the open-cast coal mine Grazweiler in the Rhineland was blocked by around 1.500 activists in August 2015.

In May 2016, Ende Gelände managed to bring together an international coalition under the slogan 'Stop Coal, Protect the Climate!'. During an action weekend, activists occupied the coal mine Welzow-Süd and the coal-fired power plant Schwarze Pumpe. Especially, the occupation of coal excavators and other mining infrastructure brought a lot of media attention. Blocking the coal mine was informed by tactics from the anti-nuclear movement. In total, both the mine and the power plant were occupied for more than 48 hours. The occupations were attended by roughly 3.500 people (Goodman *et al.* 2020, pp. 147–148). This was the first time in German history that a coal-fired power station had to reduce its output (Toewe 2017, p. 92). Ende Gelände spokesperson Insa Vries explained the success both by looking at the numbers of people in the protest as well as the general social acceptance of demonstrators (ibid., pp. 92–93).

Direct action is especially informed by trust. Some groups within Ende Gelände have engaged together in direct action for a long time (Ende Gelände 2020a, p. 3). The mobilising tactic is around 'fingers'; groups of people who are responsible for a common task such as blocking railways or mine infrastructure (Ende Gelände 2020c, p. 4). From 2015 onwards, Ende Gelände managed to organise at least one blockade per year, two in 2019. The press mostly picks up on blockades and reports about the movement's goals when the movement clashes with police forces.

Educating

In order to increase support, Ende Gelände invests a lot of time on education and reflexive practices. The primary educational goal is to show that coal extraction is 'inefficient and dirty' (Ende Gelände 2017). Some of the arguments made against coal are as follows:

- Lignite is the most CO_2-intensive source to generate energy.
- Even in the most modern coal-fired power stations, half of the energy gets lost in the production process.
- Over-capacity from lignite crowds out renewable sources of energy.
- Direct and indirect subsidies for coal would amount to 4.5 billion Euros per year.
- Energy generation from coal is structurally incentivised to feed the grid.
- Renewable energy from wind and solar are cheaper than new coal-fired power plants.
- Even new coal-fired power plants cause health hazards and might cause asthma, heart attacks, and lung cancer.
- Ende Gelände also claims that decentralised power generation from renewables coupled with investment in R&D for storage brings more energy security than fossil-fuel production.

Educational aspects are not only about making the argument against coal in leaflets or blogposts, learning also takes place within the movement. Ethnic minorities voiced their dissatisfaction with the lack of ethnic diversity, especially in the leadership group. The failure to reflect on the special needs of ethnic minorities in environmental struggles was decried. For example, it has been reflected that undocumented migrants cannot risk arrest like German activists (Ende Gelände 2020b). The movement thus aspires to a form of auto-critique in order to decolonise their own movement practices.

Ende Gelände also runs a podcast to discuss forms of direct action or other aspects of past and future protest action. Film screenings are part of the educational programme of Ende Gelände. An international selection of movies is shown at climate camps and at information evenings in Berlin and other German cities. Film screenings are seen as effective educational tools understanding broader aspects of anti-fossil struggles, but also imagining ways of how to connect to other struggles.

Connecting

According to Goodman *et al.* (2020, p. 148), Ende Gelände manages 'to shed light on the regional problem but to frame it in a global perspective'. The movement is eager to link their struggle to wider movement struggles both nationally and internationally. Connecting the anti-coal movement to other struggles forms an explicit policy of Ende Gelände.

> We also find it extremely important to reach out to other movements. The climate crisis is not an ecological crisis, but it is interwoven with various power relations. Those different social, political, cultural and economic power relations are inter-dependent and mutually reinforcing.
>
> (Ende Gelände 2020c, p. 9)

Ende Gelände teams up with the movement Alle Dörfer Bleiben (all villages remain) with a support base in the villages endangered by coal companies. All Villages Remain targets big energy corporations using lignite.

> For people in more than a dozen villages in Germany, the future is at stake. Houses, churches, forests and fertile farmland are to be consumed by open cast lignite mines. This is happening despite it being clear that renewable energy production is possible and a coal exit [is] absolutely necessary to stay within the 1,5°-limit. The villages are to be destroyed and 1.500 people to be displaced solely for the profits of the big energy corporations RWE, LEAG and MIBRAG, which run the open cast mines in the three German lignite mining regions.
>
> (All Villages Remain n.d.)

All Villages Remain's video material shows excavators destroying German villages contrasted by demonstrators taking village streets and occupying coal infrastructure. Their video material shows activists holding Ende Gelände and Antifa banners and flags next to yellow x's that stand for All Villages Remain.

In the Lusatia area, Ende Gelände managed to rally support among a number of residents and local decision-makers such as majors. However, alliance-building is limited by diverging interests between the movement and other groups, for example, the workers' union that organises coal workers. In 2016, Ende Gelände explicitly invited them to join the climate camp which was declined by the union (Toewe 2017, pp. 93–94). When Ende Gelände demonstrated the coal compromise in 2019, a workers' union leader criticised the movement for not respecting democratic procedure (Meisner 2019).

Prefiguring

As a short reminder, prefigurative politics basically means that social movements practice in small what they want to see scaled-up in the wider world. Prefigurative politics is particularly present at climate camps that Ende Gelände co-organises. The climate camps typically have some important educational elements such as visits to coal infrastructure and action training (Lausitzcamp 2016, p. 18). However, the most important pillar of the Lusatia climate camp has been to live alternatives by putting in practice 'a direct democratic self-management system' (Lausitzcamp 2016, p. 3). The 2016 climate camp that took place just outside the opencast coal mine Welzow-Süd, organisers imagined life in the climate camp in the following terms:

> We will live a resource conserving and direct democratic lifestyle. We know about alternatives and demonstrate their feasibility. A responsible approach to nature, awareness and social inter-personal dealing (sic.) is the basic principle and a vivid expression of our vitality and should be self-evident.
>
> (Lausitzcamp 2016, p. 4)

Direct democracy and respect for nature are two key elements of camp self-organising. Every morning, a plenum discussion with delegates from every barrio (neighbourhood) of the camp is taking place. Decisions that are taken by delegates will be discussed in barrio meetings afterwards. Awareness and attentiveness are also an integral part elements of the climate camp. Organisers do not want to 'tolerate any form of racist, anti-Semitic, sexist or homophobic language, actions or behaviour. Violence, harassment or sexual assaults will result in expulsion from the camp' (Lausitzcamp 2016, p. 10).

Conclusion: Germany's fossil-fuel frontier

The discussion first looked at coal mining in Eastern and Western Germany. Lignite mining has a long history in both German states making unified Germany the single biggest CO_2 emitter in Europe. However, it was only until 2015 that lignite mining was explicitly contested in Germany by sustained social movement action. A number of different influences from diverse social movements consolidated a number of frames and tactics to oppose lignite extraction in Germany.

Like the Fridays for Future movement, Ende Gelände is opposed to the German coal consensus and mobilises against the coal phase-out dated for 2038. However, because of its anti-capitalist framing and tactics, Ende Gelände belongs more to the radical fringe of the fossil-free movement. While the Fridays for Future movement is closer to institutional channels such as political parties and not explicitly anti-systemic, Ende Gelände considers fossil fuels integral to capitalist accumulation. As shown earlier, the German climate movement agrees to rally around the 1.5-degree goal as a demand to enter a post-fossil fuel age. However, it seems that the 1.5-degree target has become almost meaningless as most governments, corporations, and climate NGOs buy into this narrative. For most states, this means that they adapt their language of adaptation and mitigation that circumvents questions around fossil fuel use (Wainwright and Mann 2020). A crucial role for fossil-free movements is thus to hold governments to account in order to stick to their own climate targets.

Common Frontlines: Fossil-Free Struggles in South Africa and Germany

This chapter first discussed the political structure that enables coal exploitation and use in Germany. It also showed significant mobilisation against the fossil-fuel industry. However, in how far can we compare the German and South African examples? Germany is a core state of the industrialised world and a trade hegemon. South Africa is a semi-peripheral country with some features of an industrialised country. However, the wealth of South Africa is distributed very unevenly and a quarter of South Africans are suffering from food poverty (World Bank 2020). Acknowledging the limits of comparing an industrialised economy to an emerging postcolonial economy, there are nevertheless a number of similar characteristics between South Africa and Germany regarding their energy policy:

- Strong domestic energy sector dominated by large energy companies.
- Both countries are consolidated democracies.
- Both share long extractivist histories.
- Strong input of domestic coal to the overall energy mix.
- Strong governmental support of the fossil-fuel sector in form of subsidies.
- Both countries have social movements that are shaping the debate on energy intellectually and are prepared to disrupt fossil fuel operations.

Sometimes, the connection between different social movements in the same country is already difficult. For example, Calla (2020) reports how Bolivian anti-extractivist struggles and anti-racist struggles despite common problems find difficulties in joint mobilisation. The potential for transnational political resistance is therefore quite limited generally.

Looking at both the movements' framing and movement tactics, we can identify more than mere familiarity between anti-fossil movements in Germany in South Africa. Ende Gelände, Frack Free South Africa, and Save our iMfolozi articulate demands that are going beyond their own backyards linking site-specific demands to wider claims around linking fossil-fuel extraction to climate change. Anti-fossil-fuel movements in Germany and South Africa are both rooted in various historical environmental movements. In South Africa, the anti-fossil-fuel movement is rooted in the conservation movement but is increasingly also informed by EJ theory and practice. In Germany, the Ende Gelände movement is inspired by some movement tactics of the anti-nuclear movement and British climate camp activism and is ideologically close to the alter-globalisation movement.

All three movements analysed started in 2015, the year of the Paris climate summit. Other than most grassroots environmentalist movements, they all managed to persist in engaging in sustained opposition against the fossil industry. In the German case, peak mobilisation was in 2019 when coal was high on the political agenda and mass mobilisation against the coal compromise took place. Ende Gelände became more and more an established political movement on the left fringe of the German environmental movement. 2019 was also the year that the Berlin intelligence services put Ende Gelände on its list signalling a potential threat to the constitution (Verfassungsschutz Berlin 2019). The German anti-fossil movement subsequently managed to increase the pressure against deforestation to access coal fields for coal extraction or the construction of a new coal-fired power plant in Germany. Every extension or construction of fossil-fuel infrastructure in Germany is therefore in danger of being blocked in the future.

Fossil-free movements in Germany and South Africa are actively engaged in energy policy from below (Temper *et al.* 2020). The most-reported actions of fossil-free movements are, in fact, when they block the routine processes of the fossil-fuel production cycle. The production cycle starts from exploration to extraction and finally fossil-fuel use. When fossil-free movements are disrupting this cycle, they become agents of energy and climate politics. By blocking and delaying, fossil-free movements are also standing outside routinised institutional politics and are provoking the state to take sides. In both the case of Germany and South Africa, we have seen how the state reacted to extra-parliamentary challenges with mild to moderate forms of coercion. In this regard, the 'sunrise industry' discourse of the South African government reveals some ambiguity. On the one hand, the mining industry and its workers are portrayed as an integral part of South Africa's development. On the other hand, opposition to mining is regarded as un-patriotic as the parliamentary debate on fracking revealed. ANC politicians showed in favour of fracking exploration as a 'game-changer', some opposition parties including the African Christian Democratic Party (ACDP) voiced their opposition. Some other parties at least indicated discomfort with the environmental risks from fracking.

Both the experience of South Africa's as well as Germany's fossil-free movements however revealed that the state is not a unified actor when it comes to fossil fuel policy. In the city of Berlin, the local governing parties actually disagreed over the mention of Ende Gelände in the Intelligence report. In the face of commitments to decarbonise the grid and the expectation to legislate for more sustainability, the scope for fossil-fuel infrastructure expansion for German legislators will be smaller in the future. However, taking into consideration the diversity of fossil fuel sources and fossil fuel imports, it is difficult for social movements to mobilise against these multiple sources. Only some of the fossil-fuel energy imports of the German state are politicised by Ende Gelände. In the face of embargos on Russian oil and gas, opposition to oil and gas imports from the US and other countries will be even harder to organise.

It will be increasingly difficult for democratic fossil-fuel-producing states in the future to make the argument in favour of fossil-fuel production considering their own commitments to CO_2 reductions. Fossil-free movements thus opened a political space rendering fossil-fuel extraction more contentious.

Fossil-free movements allocate a lot of time to other movement tactics that are less visible to the general public. A key tactic is providing education on the environmental damages coming from fossil-fuel extraction and burning. Producing material to counter claims around sustainable fossil fuel use is particularly important in South Africa where fracking is still largely unknown. Corporate reporting from institutions like Bundesverband Braunkohle e.V. or Rhino Oil and Gas on the benefits of fossil-fuel extraction and use are met with counter-reports such as the 'Fracking in South Africa: A beginners guide' or the anti-coal factsheets of Ende Gelände (2017). The movements are thus active in both concrete acts of opposition through blocking and delaying fossil-fuel infrastructure, but also engage in the battle of ideas around energy supply.

Fossil-free movements in South Africa and Germany are engaging in prefigurative politics. FFSA members started to declare 'Frack Free zones' in their neighbourhoods and advertised for a transition to renewable energies. Ende Gelände in Germany organised climate campsites just outside coal mines where they put in practice 'a direct democratic self-management system' (Lausitzcamp 2016, p. 3).

Both in South Africa and in Germany, the racial composition of protesters is rather homogenous. There are some efforts to make the movements more inclusive and multi-ethnic to become more representative of the respective societies at large. There are some emerging practices of auto-critique for these movements to reflect on ways to make the fossil-free movement more diverse. Let us now finally look at the potentials for global mobilisations against fossil fuels. In this context, I will particularly look at the potentials of different action frames to organise against the hegemony of fossil-fuel exploitation and use.

Global Struggles at the Fossil-Fuel Frontier

Probably the most promising global campaign to stop fossil fuels was the Ecuadorian Yasuní-ITT campaign against oil drilling under the *Parque Nacional Yasuní*, a massive sink for CO_2 emissions and home to the most diverse insect, tree and some animal species (Sovacool and Scarpaci 2016). The Ecuadorian government's proposal backed by indigenous groups and social movements was to raise around half the crude oil worth of revenue not to drill the field. The rationale was that 'more than 400 million tons of carbon dioxide' could be 'sunk' in the project – leaving the oil asset stranded in the soil (Sovacool and Scarpaci 2016, p. 159). However, there was a considerable gap between donations by the international community and funds raised as only around half a percent of the target amount had been pledged in a six-year period (ibid., p. 156). Finally, the campaign lapsed in 2013 and former

Bolivian president Correa shelved the idea and drilling started in 2016. Sovacool and Scarpaci (2016) identified the major flaw of the project in the lack of political will on the side of international governments fearing that the Yasuní-ITT campaign would set a precedent for them to pay more for unburnt fossil fuels in the future.

The conclusion from this initiative for climate movements is sobering. However, and importantly, as seen in the preceding chapters on fossil-free social movement action in South Africa and Germany, social movements are raising the bar for governments to pursue a fossil-fuel-heavy policy path. For social movements, it shows that governments are not necessarily and uniformly interested in entrenching fossil-fuel extraction. We can expect that the state will increasingly be pressured to take a stance on individual fossil-fuel projects and to move away from fossil-fuel interests. From a government-perspective, resource governance of the future will likely be characterised by twin-risks from both climate feedback-loops from excessive neo-extractivism as well as militant social movement action. The fight against fossil fuels is going to be against climate inert governments and corporations that are fighting back against climate legislation. From a social movement perspective, it will be of importance to make strategic decisions about allies. In recent mobilisations against fossil fuels, the majority of claims were directed against governments. The conclusion from this stand-off should not be to stay clear from institutional politics and political power. Challenges to climate change inertia should come from inside and outside parliaments. Especially in countries that are dangerous places for environmental activists, getting seats in parliament give elected activists the opportunity to speak truth to power. Parliamentary representation also ensures that resources can be directed at mobilising for a green agenda.

Even though the challenge to contain the fossil-fuel industry seems daunting and the time to act short, we should not forget that fossil-free movements are a very recent phenomenon. Targeting fossil-fuel infrastructure only recently became a repertoire of contention. Already, they make an impact in changing the debate and delay or stop the construction of fossil-fuel infrastructure (Temper *et al*. 2020). In fighting fossil-fuel interests, they render visible what I termed the fossil-fuel frontier. As seen in this book, the frontier always has local and global dimensions. The impacts of fossil-fuel burning are global. From the perspective of social movements, it therefore makes sense to highlight that these struggles are part of the same quest for a post-extractivist and post-fossil-fuel agenda. Especially for smaller movements concerned with the everyday struggle to mobilise resources, it is difficult to look for allies elsewhere. The case of anti-fracking activism in South Africa has shown that activists are sometimes left quite isolated. Communal experiences such as the Frack Free Festival in Matatiele are vital to connect the dots between activists mobilising for the same cause. Activist leaders are instrumental in bringing together activists and supporters under the same banner at places like the Frack Free Festival. Growing the movement and scaling-up protest are the main challenges for fossil-free movements. With all major fossil-fuel-producing states not conforming to their own emission targets, this challenge of fossil-free movements could not be more pressing. Therefore, I will finally look at how frames and tactics could be aligned towards to challenge persistent fossil-fuel hegemony globally.

Global Social Movement Framing

A key question arises: What could unite fossil-fuel movements around the globe? What kind of master-signifier could connect the dots between different site-specific struggles against fossil fuels? Finding a common language is a key to move in the same direction. This is true for the global fossil-free movement but also in their relation to other environmental movements in both the Global South and North. In what follows, I briefly want to discuss the affordances and limits of frames bringing a common agenda to life that has ending fossil fuels at heart (see table 7.1), but also reflect a more widely alternative socio-ecological horizon. I discuss the Green New Deal, de-growth and EJ frames as three of the most recurring global environmental frames. All three frames are aimed at ending global fossil-fuel hegemony, but each focuses on slightly different aspects of injustices and change strategies. I conclude that all have the potentials to bring together movements from around the globe, but the Green New Deal and de-growth frames are limited by a Global North bias.

Table 7.1 Main movement environmental frames and their potential for global claim-making.

	Green New Deal	*De-Growth*	*Environmental Justice*
Main idea	Policy proposals and investment plans around energy, public transport, green jobs among others	Reduction of material size of the economy, redistribution within planetary boundaries	Disproportionate harm from industrialisation in the Global South and black, indigenous, people of colour communities (BPoCs); related concepts: environmental racism, energy justice
Main actors	Social movements, progressive politicians and political parties in the USA and Europe	Rooted in European social movements	Global social movement networks and local NGOs in the Global South and North
North–South relationship	Focus mostly on economic recovery in Global North; very little resonance in the South	The Global South perspective is taken into consideration, but scepticism and little resonance among Southern movements	Some elements of concerted claim-making between the North and South

Source: Author's depiction

Green New Deal

Green New Deal (GND) proposals are ambitious policy plans ranging from energy, public transportation, green jobs, fair wages, debt relief, public investment in infrastructure, and other proposals that combine social and ecological stimuli. The strength of GND proposals is that they shift the debate to a change agenda rather than merely being an elaborate critique of the status quo (Smith 2021; Riexinger 2020; Klein 2019). Crucially, there is also some reflection on the relationship between the Global North and the Global South and discussion on what a global Green New Deal could bring to make up for rampant inequality globally (Lenferna 2020; Varoufakis and Adler 2019). However, debates and proposals for a GND mostly come from the Global North and currently the case for green recovery amid the Covid-19 crisis is having the recovery of the Global North in mind.

In the USA, the Sunrise movement demands a legislated GND that guarantees 'no government investments, bailouts or subsidies may go to support fossil fuel polluters or the expansion of fossil-fuel infrastructure at home or abroad' (Maunus 2021). Sunrise also demands that 'fossil fuel executives and other agents of corruption must be brought to trial by Congress and the Biden administration' (ibid.). Before the election of Joe Biden to the White House Naomi Klein commented on the need for strong social movement action for any meaningful GND legislation: 'any administration attempting to implement a Green New Deal will need powerful social movements backing them up and pushing them to do more' (Klein 2019, p. 261). Furthermore, Klein demands a holistic GND

> [E]xplicit about keeping carbon in the ground, about the central role of the US military in driving up emissions, about nuclear and coal never being 'clean', and about the debts, wealthy countries like the United States and powerful corporations like Shell and Exxon owe to poorer nations that are coping with the impacts of crises they did almost nothing to create.
>
> (Klein 2019, p. 264)

It seems quite unrealistic to overthrow the US military–industrial complex, revolutionise the grid, and usher in debt relief to the poorer nations. However, the rhetoric of making the impossible possible is shared by Extinction Rebellion saying that 'every crisis contains the possibility of transformation' (Extinction Rebellion n.d.).

In Europe, the last years have also seen a proliferation of GND initiatives from the British Labour Party to the German Green Party and Left Party to the pan-European Democracy in Europe Movement 2025 (DiEM25) (Smith 2021; Riexinger 2020; Grüne Hessen 2009). In the eyes of Smith (2021) however, all initiatives save for the DiEM25's GND proposal fall short in putting enough emphasis on resource overuse and growth ambitions that would crank up emissions. The GND is not yet a consolidated policy proposal but used by a number of organisations and parties to mean different things. Perhaps more important in this context is that the GND has not seen much resonance in fossil-free grassroots movements. Hesitation to take up the GND narrative is perhaps at least partly due to the European Commission's similar use of terminology in their flagship proposal for a European Green Deal.

De-Growth

The global de-growth movement takes its cues from the 1972 publication of *Limits to Growth* and the Club of Rome's subsequent warnings of global overshoot from resource use, consumerism, and population growth (Meadows *et al.* 2004). In the early 2000s, in France and other European countries, the slogan décoissance' (French for de-growth) had some mobilising potency that was used to protest car use, over-consumption and advertisement (Kallis *et al.* 2018, p. 292). Renewed critiques of the growth paradigm have attracted a number of widely noted publications (e.g. Hickel 2020; Kallis 2019; Jackson 2017; Raworth 2018). The Leipzig de-growth conference drew some 4.000 participants (Kallis *et al.* 2018, p. 292). De-growth can partly be seen as a critique of growth-oriented GND proposals as de-growth proponents have doubts about economic growth in general – including green growth (Hofferberth and Schmelzer 2019). Arguments around absolute or relative decoupling of growth from emissions are proven to be unrealistic even considering rapid technological innovation. According to Hickel and Kallis (2020), there is no way to grow the economy out of the ecological crisis. Rather, green growth would entrench problems around resource over use.

From a movement perspective, de-growth insights have informed a number of mobilisations and shaped movements' demands including decommodification, decentralisation, and post-extractivism (Burkhart *et al.* 2020; Brand 2015). More than being an organised movement, de-growth is rather a loose network of mostly scholars (denkhausbremen 2018). There are a number of problems identified by activists with the notion of de-growth outside the Global North. The reasons why de-growth is less appealing in the Global South context include fundamentally different political realities (where 'growing' is not necessarily seen as bad), its Western-centric approach (eurocentrism) and perceived lack of transformative ambition (Rogríguez-Labajos *et al.* 2019, p. 177). From a South African movement perspective, de-growth is a rather neglected perspective similar to GND perspectives. However, as we will see, both global GND proponents (rather than the ones limiting their ambition to individual countries) and de-growth activists have some shared ambitions for global EJ.

Environmental Justice

EJ is a concept that emerged from environmental struggles provoked by toxic waste dumping in African-American communities in the 1980s (Martinez-Alier *et al.* 2016, p. 732). Back in the 1980s, the call for EJ was informed by the greater likelihood of African-Americans, Native Americans, Asian-Americans, and Latinos living near hazardous and polluting infrastructure (Martinez-Alier 2002, p. 170). The term thus brings together the study of race relations and environmental injustices also referred to as environmental racism (Martinez-Alier 2002, p. 168). The EJ narrative transgresses the conservation-centred approach to environmentalism as it takes into view a whole set of issues that are not considered by the conservation frame. According to eminent EJ scholar Martinez-Alier, the US-EJ movement was successful in scaling up concerns for the environment.

> [the US-EJ movement] shifted the whole discussion about environmentalism in the USA away from preservation and conservation of Nature towards social justice, it destroyed the NIMBY image of grassroots environmental protests by turning them into NIABY protests (not in anyone's backyard), and it expanded the circle of people involved in environmental policy.
>
> (Martinez-Alier 2002, p. 173)

Other than the de-growth and GND narratives, EJ framing has been frequently used in the history of environmental struggles in South Africa. The EJ narrative is used both to raise awareness for unrehabilitated dumpsites from apartheid as well as post-apartheid environmental injustices (Martinez-Alier 2002, p. 181). In 2004, the Durban Group for Climate Justice has been an attempt to institutionalise different kinds of mobilisations around these issues (Bond 2010, p. 50). Crucially, the EJ perspective takes into view imbalances between the North and South when it comes to trade relations and externalisation of pollution to the Global South (Bond 2010, p. 52; Martinez-Alier 1997). The EJ perspective is thus helpful to oppose new commodity scrambles in Africa (Munnik 2007). Munnik identifies growing pressures on commodity frontiers in Southern African and the imperative for movements to work together under the EJ umbrella.

> The balance of political power in all of our societies, while dynamic and subject to ongoing change, suggests that Southern Africa will face increasing environmental injustice in the way its resources are used, including the ongoing enclosure of people's commonly owned and used resources into private domains, the unequal and unfair relationships between local populations, national decision-makers and private investors, the ongoing exclusion from [the] decision-making of local communities, and the intensifying imposition of externalities. Current developments, specifically the commodities boom and the rapid expansion of South African business and industry into the region, make it increasingly less feasible for environmental justice activists in the region to continue working in isolation in our respective countries.
>
> (Munnik 2007, p. 2)

Even before the end of apartheid international conferences were held where the potentialities of the EJ concept were explored with civil society actors on the forefront (Munnik 2007, p. 3). Considering the EJ history in the US-civil rights movement, the South African experience with EJ, the narrative is most compatible with social movement action.

Comparing the scope and limitations of GND, de-growth, and EJ ambitions, we see some common ground in terms of their respective transformative ambitions. All three share some overlaps. Both GND and de-growth movements share that they want to change the global economy to make it more equitable and sustainable. While recognising global inequalities, there is little emphasis on the needs and grievances of environmental movements in the Global South. There is little resonance from the Global South movement to de-growth narratives (Rogríguez-Labajos *et al*. 2019) and GND proposals. As seen, EJ framing bears the most potential to bridge demands and grievances because of its particular roots from the struggles of oppressed and underprivileged people. For the given reasons, fossil-free struggles are likely to lose Southern movements if they embrace GND and de-growth frames as their main framing. As seen from the Save our iMfolozi Wilderness campaign, demands for EJ have been taken up to frame the protest. EJ is also compatible with other organisations claims and demands such as African unions and NGOs (IndustrALL Union 2021).

Global Social Movements Tactics

From the previous section, we saw that the EJ framing has some potential to make movement claims between Northern and Southern movements. Concerted global movement action against fossil fuels is difficult logistically and needs a well-connected and sustained group of social movements. In this section, I discuss some tactics that existing fossil-free groups use globally.

Looking at global organisations, including NGOs and think tanks (see Table 7.2), we see that a number of fossil-free movements were founded the year that the Paris Climate Agreement was agreed and in the following years. All of these groups are pointing towards the incompatibility of Paris objectives and the deep global entrenchment of fossil fuels. These movements are more and more prepared to engage in blocking tactics. Blockades are among the most used tactics in global EJ mobilisations (Martinez-Alier 2016, p. 438). Previous to that global fossil-free organisations were mostly education-centred and connecting to form transnational networks of exchanges to shift the parameters of the climate change debate to fossil fuels. The biggest concerted action of the fossil-free movement took place in May 2016 when social movements on five continents took joint action blocking fossil-fuel infrastructure including the biggest coal export facility in Newcastle, Australia where 2.000 people blocked the harbour and the train lines as well the shutdown of the Flos-y-Fran opencast coal mine in Wales for twelve hours (Break Free 2016). May 2016 also saw the aforementioned blockade of the German coal mine

Table 7.2 International anti-fossil organisations and tactics of anti-fossil organisations (selection).

Organisation	Organisation type/ URL	Active since	Main tactic	Size	Regional focus
Shale must fall	Campaign network https://shalemustfall.org	2020	Connecting	Big	Global
Fossil fuel non-proliferation treaty	Campaign network https://fossilfueltreaty.org	2019	Educating and connecting	Medium	Global/ UK-focused
Fridays for Future	Social movement https://fridaysforfuture.org	2018	Lobbying and educating	Big	Global
Extinction Rebellion	Social movement https://rebellion.global	2018	Blocking and educating	Big	Global
Sunrise Movement	Social movement www.sunrisemovement.org/	2017	Blocking and educating	Big	USA
Ende Gelände International	Social Movement www.ende-gelaende.org/en/	2015	Connecting and blocking	Big	Germany-focused
Environmental Justice Atlas	Research network https://ejatlas.org/featured/blockadia	2015	Educating and connecting	Big	Global
DivestInvest	Activist Network www.divestinvest.org/about/	2014	Diverting finance	Big	Global
Leave it in the ground (LINGO)	NGO www.leave-it-in-the-ground.org/	2011	Lobbying and educating	Small	Europe
Carbon Tracker	Think tank https://carbontracker.org	2011	Educating	Big	Global
350.org	NGO www.350.org https://gofossilfree.org/	2008	Movement building and educating	Big	USA
Oil Change International	NGO http://priceofoil.org/	2005	Lobbying and educating	Big	USA

Source: Author's compilation

Welzow-Süd and the coal-fired power station Schwarze Pumpe, the occupation of train tracks in Anacortes in the state of Washington at the March Point oil refinery, actions against the Pecém power plant in Fortaleza, Brazil and direct action in Aliağa, Turkey against a coal waste site and new fossil fuel plant projects.

The vast majority of movements continue to use non-violent tactics (Malm 2021). While the Fridays for Future movement does not block fossil-fuel infrastructure directly, they block inner cities to show their opposition to industrial policy that is not in line with the goals formulated in the Paris agreement. This is a means of drawing a line symbolically and raising awareness of people living in inner cities. Globally movements are educating activists to become leaders in their struggle against fossil fuels. Larger social movements such as the Sunrise Movement and Extinction Rebellion are dedicating some of their resources to train new leaders and are quite meticulous when it comes to action planning. A key challenge for global fossil-free action is how to render visible the common but differentiated struggles. Organisations such as the EJAtlas visualise these common struggles analysing that 'conflicts usually arise from structural inequalities of income and power' (EJAtlas, n.d.). Similarly, the Shale Must Fall initiative wants to 'increase the global visibility of the frontline struggles from the Global South and other extraction zones' (Shale Must Fall 2020).

As reported earlier, a number of these struggles organised some spectacular global actions that have been widely recognised to be part of a global insurgence to 'keep it in the ground'. Currently, we see a global slump of social movement action more generally and fossil-fuel-free activism in particular. Global lockdowns and tight restrictions on the freedom of assembly made it increasingly difficult for movements to uphold the pressure and bring the numbers to the streets (Pleyers 2020). Also, the agenda has shifted to focus on the health sector and smaller acts of solidarity in the neighbourhoods. It will be a key for fossil-free movements to cease windows of opportunities to challenge the fossil-fuel industry that arise from devastating extreme weather events. A crucial element for mobilisation will be to be wary of the fact that the problem is truly global. The focus of most fossil-free organisation is still on the Global North. However, more and more spaces of fossil-fuel extraction are situated in the Global South with governments of the North slowly de-carbonising their energy and transportation systems. The challenge of the future will increasingly compensate for the unburnt fossil fuels in developing countries in the Global South. Unless a broad civil society coalition will ensure that compensation payments for unburnt deposits will flow from the North, most developing states will exploit fossil fuel deposits well into the future to cater for their developmental needs. Fighting fossil fuels will not be enough. The challenge will be to formulate political alternatives in the energy sector and beyond. A global transition towards renewable energy needs to follow democratic protocol; otherwise, people will feel left behind much like mine-affected communities do now. A new sustainable and renewable hegemony needs to find inclusive ways to ensure bottom-up democracy that can be scaled-up.

Note

1 An example is the former Prime Minister of Saxony Stanislaw Tillich (Christian Democrats) who became chairman of the supervisory board of a lignite company Mitteldeutsche Braunkohlegesellschaft (Götze and Joeres 2020, p. 187). As Götze and Joeres

show, influential networks within the CDU are open for climate change denialism and lobby against laws to halt climate change.

References

All Villages Remain, n.d. *Welcome.* Available from: www.alle-doerfer-bleiben.de/english/

Bond, P., 2010. Climate justice politics across space and scale. *Human Geography*, 3 (2), 49–62. doi:10.1177/194277861000300204

Brand, U., 2015. *Degrowth und post-Extraktivismus: Zwei Seiten einer Medaille?* (Working Paper 5/2016). Jena: DFG-Kolleg Postwachstumsgesellschaften.

Break Free, 2016. *On six continents, thousands of people took bold action to break free from fossil fuels.* Available from: https://breakfree2016.org

Brown, G., Feigenbaum, A., Frenzel, F. and McCurdy, P., 2018. *Protest camps in international context: spaces, infrastructures and media of resistance.* Bristol: Policy Press.

Bundesanstalt für Geowissenschaften und Rohstoffe, 2020. *Deutschland – Rohstoffsituation 2019.* Hannover. Available from: www.deutsche-rohstoffagentur.de/DE/Themen/Min_rohstoffe/Downloads/rohsit-2019.pdf;jsessionid=FA0C37EF58200AE353A9A4B6CA0EB589.2_cid284?__blob=publicationFile&v=5

Bundesregierung, 2020. *Ending coal-generated power.* Available from: www.bundesregierung.de/breg-en/news/kohleausstiegsgesetz-1717014

Bundesverband Braunkohle e.V., n.d. *Übersicht und Geschichte der Reviere.* Available from: https://braunkohle.de/braunkohle-in-deutschland/uebersicht-und-geschichte-der-reviere/

Burkhart, C., Schmelzer, M. and Treu, N., eds., 2020. *Degrowth in movement(s): exploring pathways for transformation.* Winchester: Zero Books.

Calla, P., 2020. The difficulties of connecting anti-extractivist and anti-racist struggles in contemporary Bolivia. *In*: J. Hooker, ed., *Black and indigenous resistance in the Americas.* Lanham: Lexington Books, 189–216.

denkhausbremen, 2018, January 23. *Nina Treu: Die Degrowth-Bewegung ist fast noch gar nicht institutionalisiert.* Available from: https://denkhausbremen.de/nina-treu/

Deutscher Bundestag, 1998, September 3. *Stenographischer Bericht 247. Sitzung.* Bonn. Available from: https://dserver.bundestag.de/btp/13/13247.pdf

Ende Gelände, 2015. *Invitation to the preparation meeting of "Ende Gelände – leave it in the ground" 1–3 May 15 near Cologne.* Available from: https://2015.ende-gelaende.org/en/node/39.html

Ende Gelände, 2017. *Ende Gelände! Kohle stoppen Klima schützen, weil . . .* Available from: www.ende-gelaende.org/wp-content/uploads/2017/08/fact_sheet.pdf

Ende Gelände, 2019. *Who shut that shit down? We shut shit down!* Available from: www.ende-gelaende.org/en/news/we-shut-shit-down/

Ende Gelände, 2020a. *Solidaritätserklärung der NGOs mit Ende Gelände nach Angriff des Verfassungsschutz.* Available from: www.ende-gelaende.org/news/klimabewegung-wir-stehen-solidarisch-zusammen/

Ende Gelände, 2020b. *Ende Gelände 2020b. Offener Brief von BPoCs an das Klimacamp und andere.* Available from: https://2020.ende-gelaende.org/wp-content/uploads/2020/09/Offener_Brief_von_BPoCs_an_das_Klimacamp_und_andere.pdf

Ende Gelände, 2020c. *Shut shit down: an activist's guide to Ende Gelände.* Hannover: Ende Gelände.

Environmental Justice Atlas, n.d. *Frequently asked questions.* Available from: https://ejatlas.org/about

Friedrich-Ebert-Stiftung, 1988. *Die Energiepolitik der DDR. Mängelverwaltung zwischen Kernkraft und Braunkohle.* Bonn: Verlag Neue Gesellschaft.

Fridays for Future, 2019, September 23. *Der größte Klimastreik der Geschichte – und das war erst der Anfang.* Available from: https://fridaysforfuture.de/ruckblick-allefuers klima1/

Fridays for Future, n.d. *Forderungen – FAQ.* Available from: https://fridaysforfuture.de/forderungen/faq/

Goodman, J., Connor, L., Kohli, K., Marshall, J.P., Menon, M., Müller, K., Morton, T., Pearse, R. and Rosewarne S., 2020. *Beyond the coal rush: a turning point for global energy and climate policy?* Cambridge: Cambridge University Press.

Götze, S. and Joeres, A., 2020. *Die Klimaschmutzlobby: Wie Politiker und Wirtschaftslenker die Zukunft unseres Planeten verkaufen.* München: Piper.

Grüne Hessen, 2009. *Mit Wums für ein besseres Europa.* Available from: www.gruene-hessen.de/partei/files/2010/05/europawahlprogramm_2009_mit_wums_fuer_ei.pdf

Häußerman, D. and Wollny, L., 2017. Anti-Kohle-Bewegung: Gegen Klimawandel, Kapitalismus und Wachstum! *In*: C. Burkhart, M. Schmelzer and N. Treu, eds., *Degrowth in Bewegung(en)*. Munich: Oekom, 34–45.

Hickel, J., 2020. *Less is more.* London: Penguin Random House.

Hickel, J. and Kallis, G., 2020. Is green growth possible? *New Political Economy*, 25 (4), 469–486. doi:10.1080/13563467.2019.1598964

Hofferberth, E. and Schmelzer, M., 2019. Gekoppelt wird ein Schuh draus: green new deal versus degrowth. *Politische Ökologie*, 159 (4), 31–38.

Illing, F., 2016. *Energiepolitik in Deutschland: Die energiepolitischen Maßnahmen der Bundesregierungen 1949–2015.* Baden-Baden: Nomos.

IndustrALL Union, 2021. *Alternative mining indaba discusses sustainable mining under Covid-19.* Available from: www.industriall-union.org/alternative-mining-indaba-discusses-sustainable-mining-under-covid-19

Jackson, T., 2017. *Prosperity without growth: foundations for the economy of tomorrow.* New York and London: Routledge.

Kahya, D., 2014, August 28. Five reasons why expanding brown coal mines might be a problem. *Unearthed Greenpeace*. Available from: https://unearthed.greenpeace.org/2014/08/28/five-reasons-expanding-brown-coal-mines-might-problem/

Kallis, G., 2019. *Limits: why Malthus was wrong and why environmentalists should care.* Palo Alto: Stanford University Press.

Kallis, G., Kostakis, V., Lange, S., Muraca, B., Paulson, S. and Schmelzer M., 2018. Research on degrowth. *Annual Review of Environment and Resources*, 43, 291–316. doi:10.1146/annurev-environ-102017-025941

Klein, N., 2019. *On fire: the burning case for a green new deal.* London: Allen Lane.

Lausitzcamp, 2016. *Info- und Programmheft.* Available from: https://archiv.lausitzcamp.de/wp-content/uploads/Lausitzcamp-2016-Programm-2016-04-18-01.pdf

Lenferna, A., 2020, April 17. The solution to the coronavirus recession is a global green new deal. *Jacobin Magazine*. Available from: https://jacobinmag.com/2020/04/coronavirus-global-green-new-deal-south-postcolonial

Malm, A., 2021. *How to blow up a pipeline.* London: Verso.

Martinez-Alier, J., 1997. Environmental justice (local and global). *Capitalism, Nature, Socialism*, 23 (1), 51–73. doi:10.1080/10455759709358725

Martinez-Alier, J., 2002. *Environmentalism of the poor: a study of ecological conflicts and valuation.* Cheltenham: Edgar Elgar.

Martinez-Alier, J., Temper, L., Del Bene, D. and Scheidel, A., 2016. Is there a global environmental justice movement? *Journal of Peasant Studies*, 43 (3), 731–755.

Maunus, L., 2021. *Green new deal year one: what we're fighting for*. Available from: www.sunrisemovement.org/movement-updates/green-new-deal-year-1/

Meadows, D., Randers, J. and Meadows, D., 2004. *Limits to growth: the 30-year update*. White River Junction: Chelsea Green.

Meisner, M., 2019, November 27. Der radikal angelegte Protest ist eine große Zumutung. *Der Tagesspiegel*. Available from: www.tagesspiegel.de/politik/dgb-vor-aktionen-von-ende-gelaende-der-radikal-angelegte-protest-ist-eine-grosse-zumutung/25270080.html

Müller, K., 2017. Heimat, Kohle, Umwelt: argumente im protest und der Befürwortung von Braunkohleförderung in der Lausitz. *Zeitschrift für Umweltpolitik und Umweltrecht*, 3, 213–228.

Munnik, V., 2007. *Solidarity for environmental justice in Southern Africa*. GroundWork Report. Available from: https://groundwork.org.za/specialreports/Solidarity%20for%20EJ%20in%20SA.pdf

Pinzler, P., 2020, July 3. Warum der Kohleausstieg so teuer ist. *Die Zeit*. Available from: www.zeit.de/wirtschaft/2020-07/kohleausstieg-energiewende-leag-rwe-entschaedigung-bundesregierung?utm_referrer=https%3A%2F%2F

Pleyers, G., 2020. The pandemic is a battlefield: social movements in the COVID-19 lockdown. *Journal of Civil Society*, 16 (4), 295–312. doi:10.1080/17448689.2020.1794398

Raworth, K., 2018. *Doughnut economics: seven ways to think like a 21st-century economist*. London: Random House.

Renn, O. and Marshall, J.P., 2016. Coal, nuclear and renewable energy policies in Germany: From the 1950s to the 'Energiewende'. *Energy Policy*, 99, 224–232. doi:10.1016/j.enpol.2016.05.004

Riexinger, B., 2020. *System change: Plädoyer für einen linken green new deal*. Hamburg: VSA.

Rinscheid, A., 2018. *Soziale Akzeptanz eines Kohleausstiegs in Deutschland und in den Kohlerevieren: Ergebnisse einer Umfrage und Conjoint-Analyse*. Hamburg: Greenpeace.

Rogríguez-Labajos, B., *et al*., 2019. Not so natural an alliance? Degrowth and environmental movements in the global south. *Ecological Economics*, 157, 175–184. doi:10.1016/j.ecolecon.2018.11.007

Sander, H., 2017. Ende gelände: Anti-Kohle-proteste in Deutschland. *Forschungsjournal Soziale Bewegungen*, 30 (1), 26–35. doi:10.1515/fjsb-2017-0004

Shale Must Fall, 2020. *Shale must fall: Clean gas is a dirty lie*. Available from: https://shalemustfall.org/2020/10/29/85/

Smith, T., 2021. Wie radikal ist der green new deal? *Prokla: Zeitschrift für kritische Sozialwissenschaft*, 51 (202), 9–30. doi:10.32387/prokla.v51i202.1928

Sovacool, B.K. and Scarpaci, J., 2016. Energy justice and the contested petroleum politics of stranded assets: policy insights from the Yasuní-ITT Initiative in Ecuador. *Energy Policy*, 95, 158–171. doi:10.1016/j.enpol.2016.04.045

Statista, 2021. *CO2-Emissionen: Größte Länder nach Anteil am weltweiten CO2-Ausstoß im Jahr 2019*. Available from: https://de.statista.com/statistik/daten/studie/179260/umfrage/die-zehn-groessten-c02-emittenten-weltweit/

Stephens, J., 2020. *Diversifying power: why we need antiracist, feminist leadership on climate and energy*. Washington: Island Press.

Stokowski, M., 2019, June 25. Warum es so viele Frauen an der Klimafront gibt. *Spiegel Online*. Available from: www.spiegel.de/kultur/gesellschaft/klimaschutz-warum-es-so-viele-frauen-an-der-klimafront-gibt-kolumne-a-1274166.html

Suhr, F., 2019, December 10. *Das sind die größten Klimasünder Europas*. Available from: https://de.statista.com/infografik/20253/unternehmen-in-der-eu-mit-den-hoechsten-co2-emissionen/

Temper, L., *et al.*, 2020. Movements shaping climate futures: a systematic mapping of protests against fossil fuel and low-carbon energy projects. *Environmental Research Letters*, 15 (12), 1–23. doi:10.1088/1748-9326/abc197

Tilly, C. and Tarrow, S., 2015. *Contentious politics*. Oxford: Oxford University Press.

Toewe, S., 2017. Ende Gelände! Hat es geschafft, handlungsfähige Akteure in einem internationalen Prozess zusammenzubinden. Interview mit Insa Vries, Sprecherin von Ende Gelände! *Forschungsjournal Soziale Bewegungen*, 30 (1), 92–95.

Uekötter, F., 2014. *The greenest nation? A new history of German Environmentalism.* Cambridge, MA: MIT Press.

Umweltbundesamt, 2020. *Bergrecht.* Available from: www.umweltbundesamt.de/themen/ nachhaltigkeit-strategien-internationales/umweltrecht/umweltschutz-im-fachrecht/ bergrecht#entwicklung-und-herausforderung-aus-sicht-des-umwelt-und-ressourcenschutzes

Varoufakis, Y. and Adler, D., 2019, April 24. It's time for nations to unite around an international green new deal [online]. *The Guardian.* Available from: https://jacobinmag. com/2020/04/coronavirus-global-green-new-deal-south-postcolonial

Verfassungsschutz Berlin, 2019. *Bericht 2019.* Berlin: Senatsverwaltung für Inneres und Sport.

Wainwright, J. and Mann, G., 2020. *Climate Leviathan: a political theory of our planetary future.* London: Verso.

World Bank, 2020. *South Africa: poverty & equity brief.* Washington, DC. Available from: https://databank.worldbank.org/data/download/poverty/33EF03BB-9722-4AE2-ABC7- AA2972D68AFE/Global_POVEQ_ZAF.pdf

Appendix

Demographics of Population in Field Sites

	Mfolozi Municipality (Somkhele, Fuleni)	uMngeni Municipality	Newcastle Municipality
Places in municipality include	Mbonambi, Ntuthunga	Cedara, Hilton, Fort Nottingham, Howick	Newcastle
Household income	No income 14.7%	12.8%	18%
	R1-R4,800 5.5%	3%	5.1%
	R4,801-R9,600 10%	5%	8.7%
	R9,601-R19,600 22.9%	18.3%	19%
	R19,601-R38,300 23.7%	20.4%	18.6%
	R38,201–76.400 12.6%	13.4%	11.1%
	R76,401-R153,800 5.8%	10.2%	8.5%
	R153,801-R307,600 2.8%	7.8%	6.5%
	R307,601-R614,400 1.5%	5.8%	3.3%
Unemployment rate/youth unemployment rate	42% overall 50.4% youth	23.9% overall 32% youth	37.4% 49%
People	98.8% black	75% black	91.9%
		1.5% coloured	0.8%
		3.8% Indian/Asian	3.2%
	0.8% white	19.4% white	3.9%
First language (excerpt)	92.5% isiZulu	61.9% isiZulu	84.9% Zulu
	1.5% isiNdebele	24.5% English	6.3% English
	2.9% English	3.6% Sesotho	3.5% Afrikaans
Education	2.2% no schooling	2.6% no schooling	2%
	44.7% some primary	38,2% some primary	40.8%
	5.9% completed primary	5.7% completed primary	5.8%
	32.9% some secondary	29,6% some secondary	33%
	13.7% completed secondary	15.5% completed secondary	15.7%
	0.6% higher education	2.2% higher education	1.8%

(*Continued*)

	Mfolozi Municipality (Somkhele, Fuleni)	*uMngeni Municipality*	*Newcastle Municipality*
Household goods	89% mobile phone	88.2% mobile phone	90.8%
	6.3% computer	27% computer	16.8%
	64.4% radio	70.8% radio	72.3%
	61.7% TV	71.3% TV	77.4%
	13.3% motor car	37% motor car	26.7%
Average household size	4.6	2.8	4.2
Facilities	5.5% flush toilet connected to sewerage	54.8% flush toilet connected to sewerage	55.8%
	7.2% weekly refuse removal	67.9% weekly refuse removal	71%
	83.7% electricity for lighting	85.5% electricity for lighting	87.2%
Land tenure	Ingonyama Trust Board (ITB); game park belongs to government; partly co-management structure with local communities in the park	Mostly privately owned land and conservation areas	Mostly privately owned land in Amajuba District; two councils fall within Ingonyama Trust Land (ITB)
Main economic activities	Agriculture (11%), tourism (11%), manufacturing (11%), trade and commerce (23%), Community Services (19%)	ecotourism, agriculture and forestry	Agriculture, mining, manufacturing

Source: Data taken from Statistics South Africa Census 2011. www.statssa.gov.za/?page_id=3839

Index

Note: Page numbers in *italics* indicate a figure and page numbers in **bold** indicate a table on the corresponding page. Page numbers followed by 'n' indicate a note.

146   *Index*

For Product Safety Concerns and Information please contact our EU
representative GPSR@taylorandfrancis.com
Taylor & Francis Verlag GmbH, Kaufingerstraße 24, 80331 München, Germany

www.ingramcontent.com/pod-product-compliance
Lightning Source LLC
Chambersburg PA
CBHW060314220326
41598CB00027B/4324

9 780367 627966